MOTOR & SENSORY PATHWAYS OF THE NERVOUS SYSTEM
An Anatomical Atlas Guide

AMY LEWIS

Motor and Sensory Pathways of the Nervous System: An Anatomical Atlas Guide
© Amy Lewis 2022

ISBN: 978-1-922717-52-8 (paperback)

Published in Australia by Amy Lewis and InHouse Publishing.
www.inhousepublishing.com.au

Printed in Australia by InHouse Print & Design.

INHOUSE PTY LTD PUBLISHING

A catalogue record for this book is available from the National Library of Australia

To my anatomy students, past and present,
and to the readers of this book,
I hope this work brings clarity on a difficult to understand topic.

CONTENTS

INTRODUCTION

The nervous system is a daunting topic to learn for many anatomy students. Students are usually in awe of the complexity of the nervous system's physiology, and how beautifully crafted the cerebrum and cerebellum are. Over the eighteen years I have taught anatomy, the topic of motor and sensory tracts, and the cranial nerve nuclei always leave students perplexed.

This book is a concise summary of the ascending sensory and descending motor tracts, and the central pathways of all twelve bilateral cranial nerves. Each motor and sensory tract is described in the form of a flow diagram which summarises the major anatomical landmarks along the pathway, either from the periphery to the cerebral cortex (sensory pathway), or from the cerebral cortex to the periphery (motor pathway). The cranial nerve central pathways are also described in the form of a flow diagram. Each flow diagram is accompanied by an anatomical cartoon illustration, which clearly shows the trajectory of the tracts and cranial nerves through the central nervous system. Each flow diagram also includes where upper and lower motor neurons synapse along the pathway, and where first, second and third order sensory neurons synapse. In each flow diagram, it is listed whether the pathway decussates, i.e., if the pathway travels ipsilaterally or contralaterally. Finally, the cranial nerve modalities are listed for each pathway of all twelve cranial nerves. It is assumed that the reader of this book has some neuroanatomy knowledge. The content of this book may be a guide for further study if you have no neuroanatomy knowledge, but want to learn.

ASCENDING SENSORY TRACTS

Lateral Spinothalamic Tract

Pain and thermal sensation

POST-CENTRAL GYRUS OF CEREBRAL CORTEX (interpretation of somatic pain) + CINGULATE GYRUS (emotional response to pain) + INSULA (interpretation of visceral pain)

↑

POSTERIOR LIMB OF INTERNAL CAPSULE / CORONA RADIATA

↑

VENTRAL POSTEROLATERAL NUCLEUS OF THE THALAMUS
- termination of A type fibres
(Synapse to 3rd order neurons)

RETICULAR FORMATION
- termination of C type fibres
(Synapse to 3rd order neurons)

↑

LATERAL SPINOTHALAMIC TRACT (CONTRALATERAL WHITE COLUMN/ ANTEROLATERAL SYSTEM/SPINAL LEMNISCUS)
(Fibres carrying pain are anterior, upper c spine nerve fibres are medial, sacral nerve fibres are lateral).

↑

FIBRES CROSS IN ANTERIOR WHITE COMMISSURES

↑

SYNAPSE ONTO CELLS OF SUBSTANTIA GELATINOSA
(In posterior grey column; synapse to 2nd order neuron)

↑

POSTEROLATERAL TRACT OF LISSAUER

↑

POSTERIOR ROOT GANGLION

↑

A TYPE SENSORY NERVE FIBRES
- fast conducting
- initial sharp pain
- thermal

C TYPE SENSORY NERVE FIBRES
- slow conducting
- prolonged burning pain
- thermal

↑

FREE PERIPHERAL SENSORY NERVE ENDINGS

Ascending Sensory Tracts

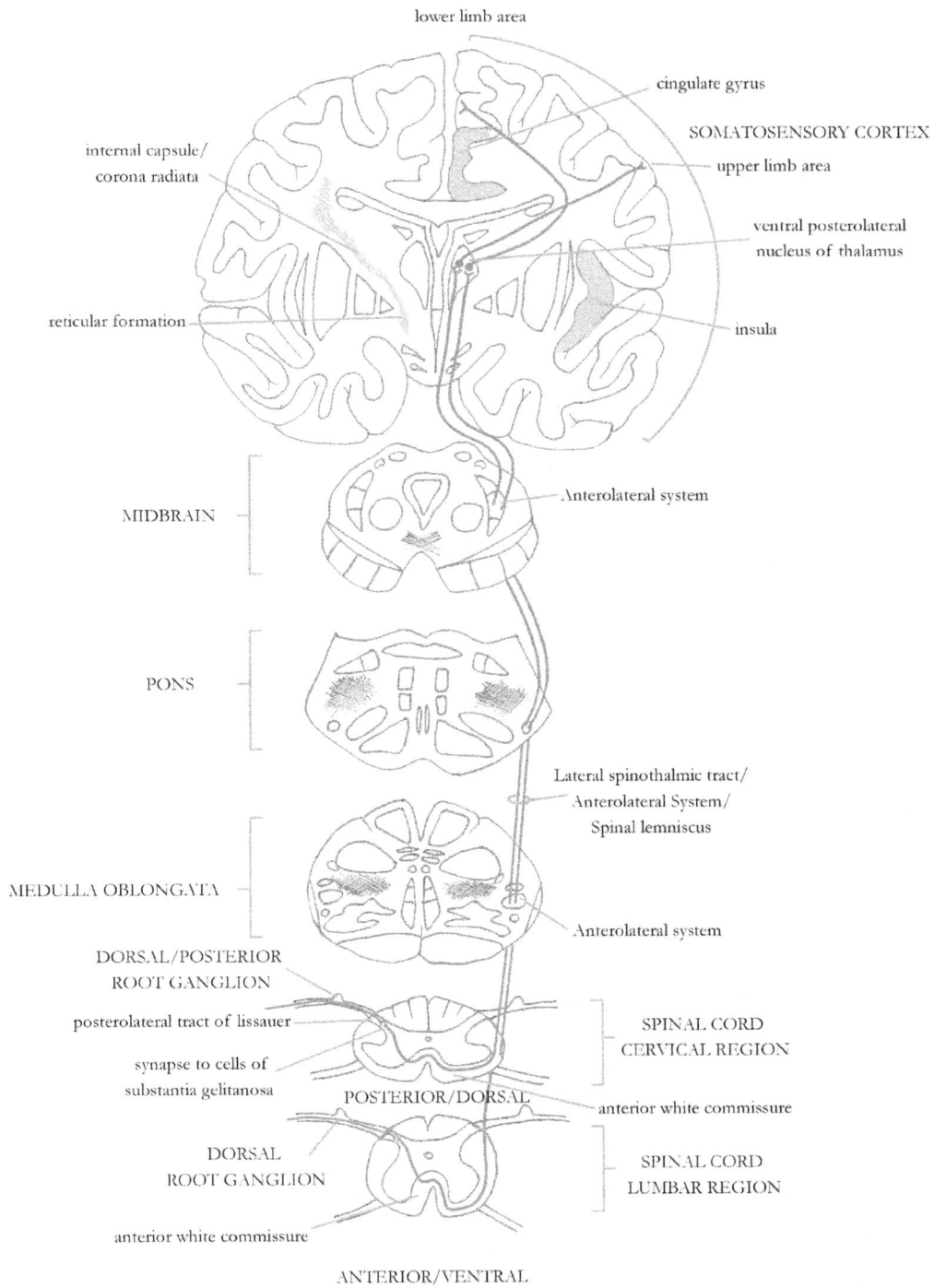

lower limb area

cingulate gyrus

SOMATOSENSORY CORTEX

internal capsule/
corona radiata

upper limb area

ventral posterolateral
nucleus of thalamus

reticular formation

insula

MIDBRAIN

Anterolateral system

PONS

Lateral spinothalmic tract/
Anterolateral System/
Spinal lemniscus

MEDULLA OBLONGATA

Anterolateral system

DORSAL/POSTERIOR
ROOT GANGLION

posterolateral tract of lissauer

SPINAL CORD
CERVICAL REGION

synapse to cells of
substantia gelitanosa

POSTERIOR/DORSAL

anterior white commissure

DORSAL
ROOT GANGLION

SPINAL CORD
LUMBAR REGION

anterior white commissure

ANTERIOR/VENTRAL

Anterior Spinothalamic Tract

Light/crude touch and pressure

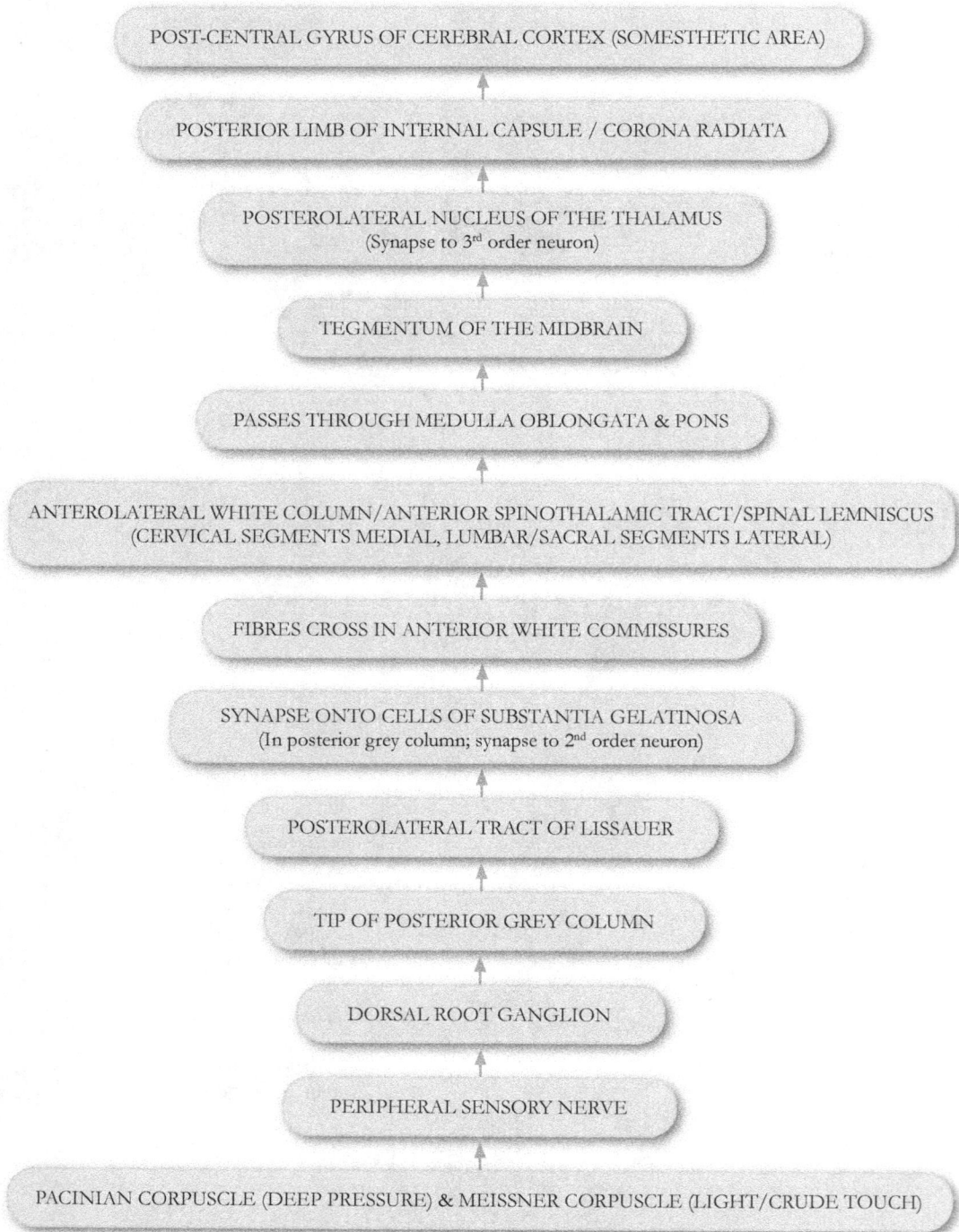

POST-CENTRAL GYRUS OF CEREBRAL CORTEX (SOMESTHETIC AREA)

↑

POSTERIOR LIMB OF INTERNAL CAPSULE / CORONA RADIATA

↑

POSTEROLATERAL NUCLEUS OF THE THALAMUS
(Synapse to 3rd order neuron)

↑

TEGMENTUM OF THE MIDBRAIN

↑

PASSES THROUGH MEDULLA OBLONGATA & PONS

↑

ANTEROLATERAL WHITE COLUMN/ANTERIOR SPINOTHALAMIC TRACT/SPINAL LEMNISCUS
(CERVICAL SEGMENTS MEDIAL, LUMBAR/SACRAL SEGMENTS LATERAL)

↑

FIBRES CROSS IN ANTERIOR WHITE COMMISSURES

↑

SYNAPSE ONTO CELLS OF SUBSTANTIA GELATINOSA
(In posterior grey column; synapse to 2nd order neuron)

↑

POSTEROLATERAL TRACT OF LISSAUER

↑

TIP OF POSTERIOR GREY COLUMN

↑

DORSAL ROOT GANGLION

↑

PERIPHERAL SENSORY NERVE

↑

PACINIAN CORPUSCLE (DEEP PRESSURE) & MEISSNER CORPUSCLE (LIGHT/CRUDE TOUCH)

Ascending Sensory Tracts

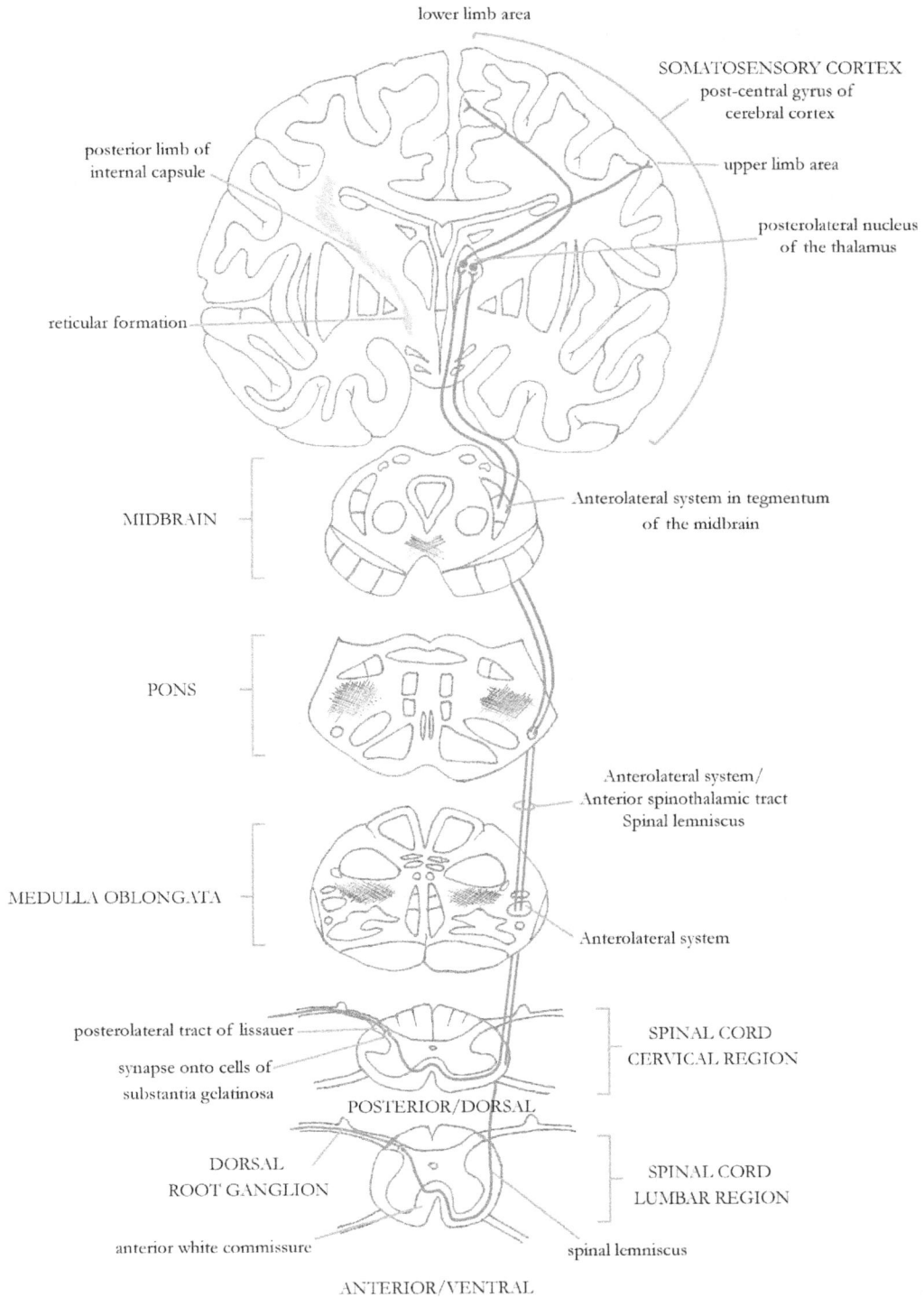

lower limb area

SOMATOSENSORY CORTEX
post-central gyrus of
cerebral cortex

posterior limb of
internal capsule

upper limb area

posterolateral nucleus
of the thalamus

reticular formation

MIDBRAIN

Anterolateral system in tegmentum
of the midbrain

PONS

Anterolateral system/
Anterior spinothalamic tract
Spinal lemniscus

MEDULLA OBLONGATA

Anterolateral system

posterolateral tract of lissauer

synapse onto cells of
substantia gelatinosa

POSTERIOR/DORSAL

SPINAL CORD
CERVICAL REGION

DORSAL
ROOT GANGLION

SPINAL CORD
LUMBAR REGION

anterior white commissure

spinal lemniscus

ANTERIOR/VENTRAL

7

Posterior/Dorsal White Column:
Fasciculus Gracilis & Fasciculus Cuneatus

Discriminative touch/vibratory sense/
conscious muscle and joint sense

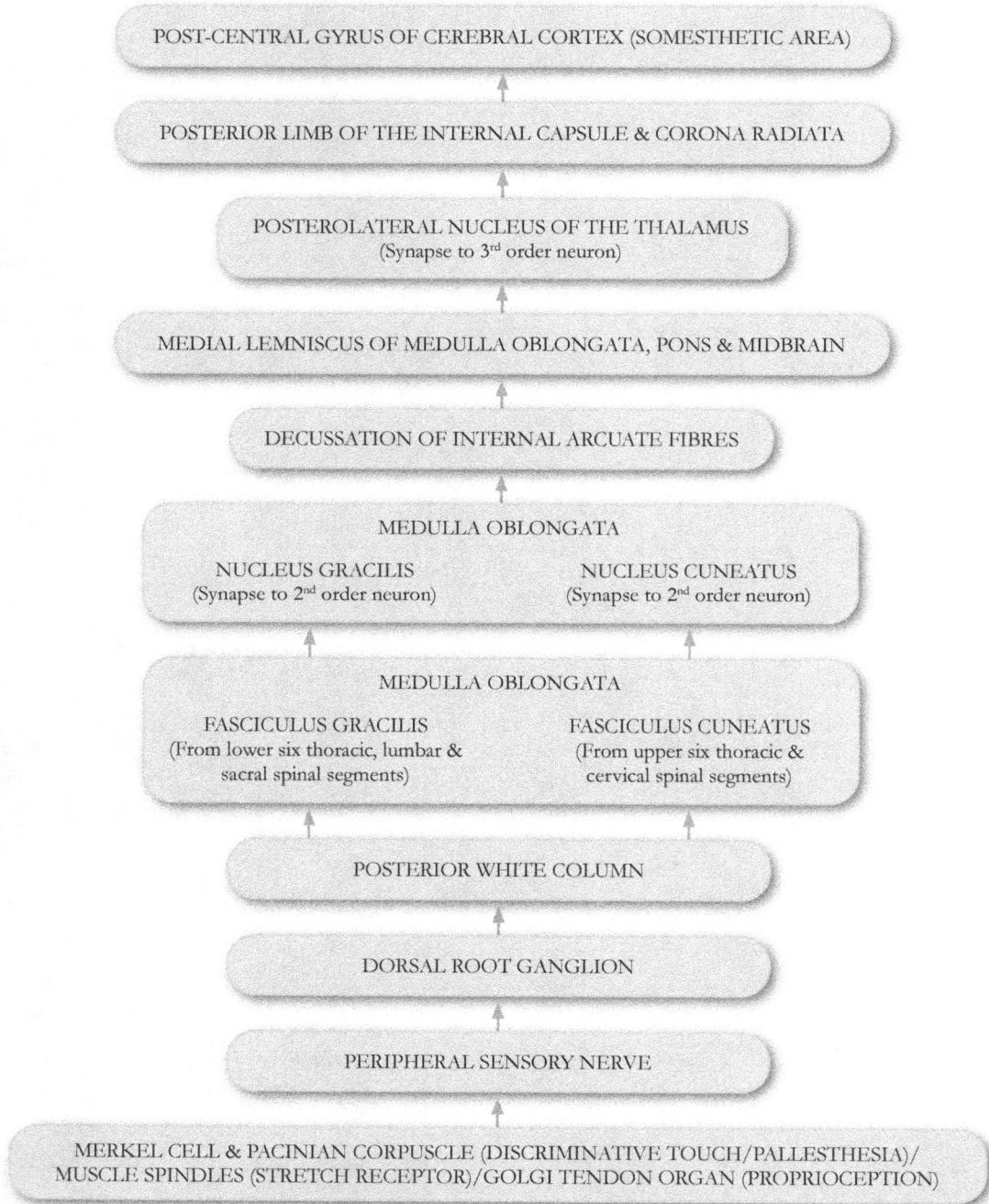

POST-CENTRAL GYRUS OF CEREBRAL CORTEX (SOMESTHETIC AREA)

↑

POSTERIOR LIMB OF THE INTERNAL CAPSULE & CORONA RADIATA

↑

POSTEROLATERAL NUCLEUS OF THE THALAMUS
(Synapse to 3rd order neuron)

↑

MEDIAL LEMNISCUS OF MEDULLA OBLONGATA, PONS & MIDBRAIN

↑

DECUSSATION OF INTERNAL ARCUATE FIBRES

↑

MEDULLA OBLONGATA

NUCLEUS GRACILIS	NUCLEUS CUNEATUS
(Synapse to 2nd order neuron)	(Synapse to 2nd order neuron)

↑ ↑

MEDULLA OBLONGATA

FASCICULUS GRACILIS	FASCICULUS CUNEATUS
(From lower six thoracic, lumbar & sacral spinal segments)	(From upper six thoracic & cervical spinal segments)

↑

POSTERIOR WHITE COLUMN

↑

DORSAL ROOT GANGLION

↑

PERIPHERAL SENSORY NERVE

↑

MERKEL CELL & PACINIAN CORPUSCLE (DISCRIMINATIVE TOUCH/PALLESTHESIA)/
MUSCLE SPINDLES (STRETCH RECEPTOR)/GOLGI TENDON ORGAN (PROPRIOCEPTION)

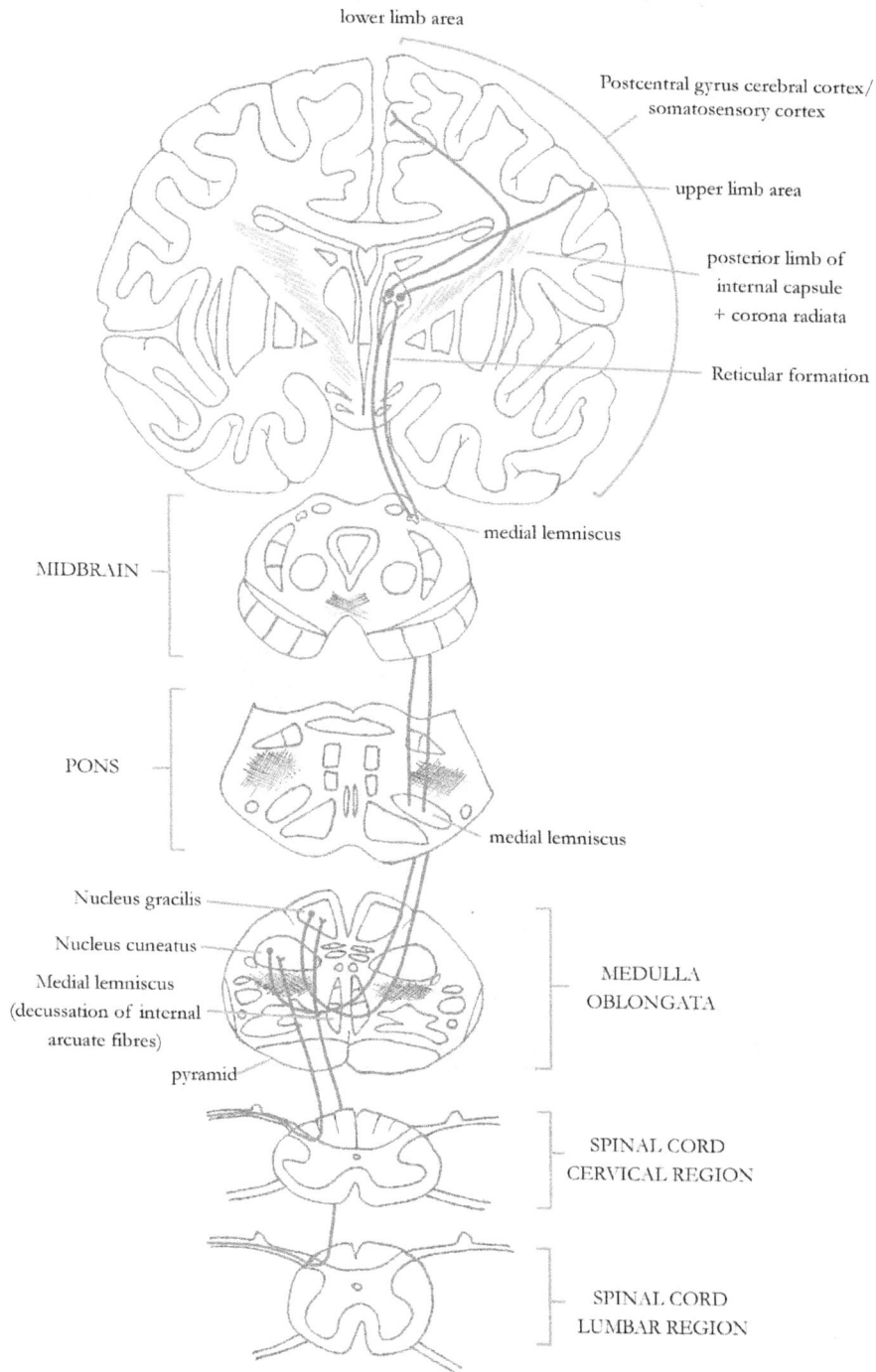

lower limb area

Postcentral gyrus cerebral cortex/
somatosensory cortex

upper limb area

posterior limb of
internal capsule
+ corona radiata

Reticular formation

medial lemniscus

MIDBRAIN

PONS

medial lemniscus

Nucleus gracilis

Nucleus cuneatus

Medial lemniscus
(decussation of internal
arcuate fibres)

pyramid

MEDULLA
OBLONGATA

SPINAL CORD
CERVICAL REGION

SPINAL CORD
LUMBAR REGION

Posterior Spinocerebellar Tract

*Muscle/joint sense pathway to the cerebellum –
unconscious proprioception for lower trunk and lower limbs*

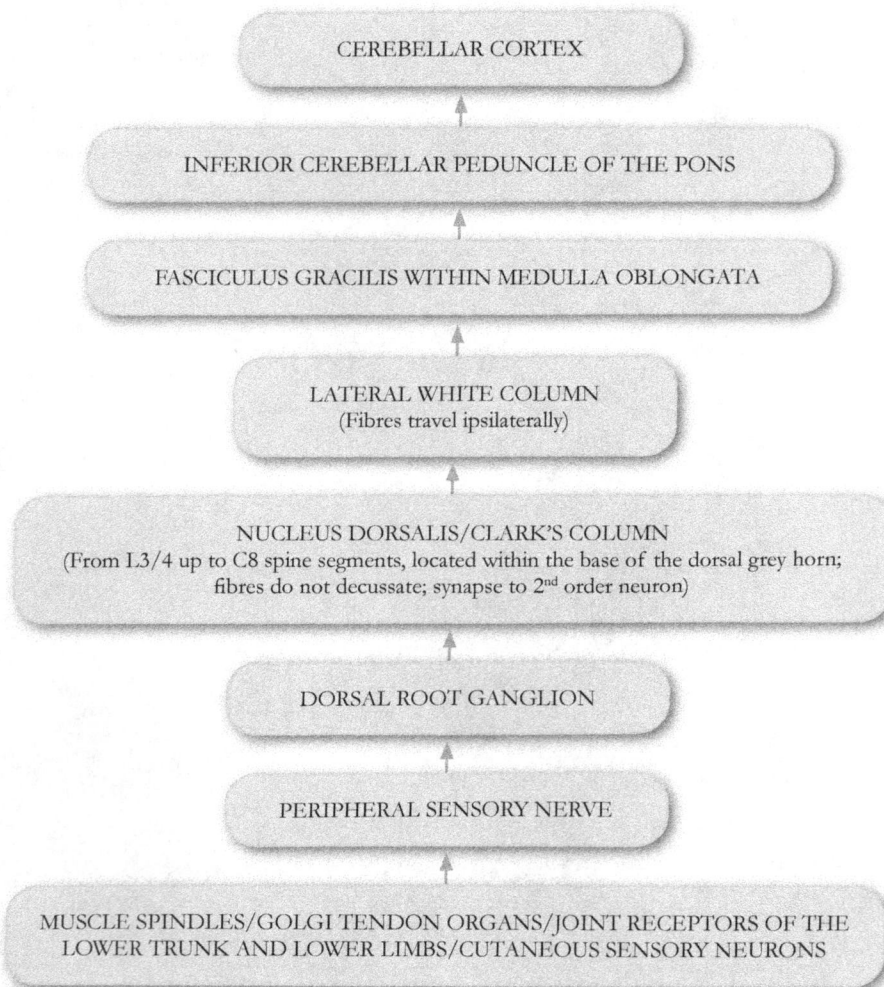

CEREBELLAR CORTEX

↑

INFERIOR CEREBELLAR PEDUNCLE OF THE PONS

↑

FASCICULUS GRACILIS WITHIN MEDULLA OBLONGATA

↑

LATERAL WHITE COLUMN
(Fibres travel ipsilaterally)

↑

NUCLEUS DORSALIS/CLARK'S COLUMN
(From L3/4 up to C8 spine segments, located within the base of the dorsal grey horn;
fibres do not decussate; synapse to 2nd order neuron)

↑

DORSAL ROOT GANGLION

↑

PERIPHERAL SENSORY NERVE

↑

MUSCLE SPINDLES/GOLGI TENDON ORGANS/JOINT RECEPTORS OF THE
LOWER TRUNK AND LOWER LIMBS/CUTANEOUS SENSORY NEURONS

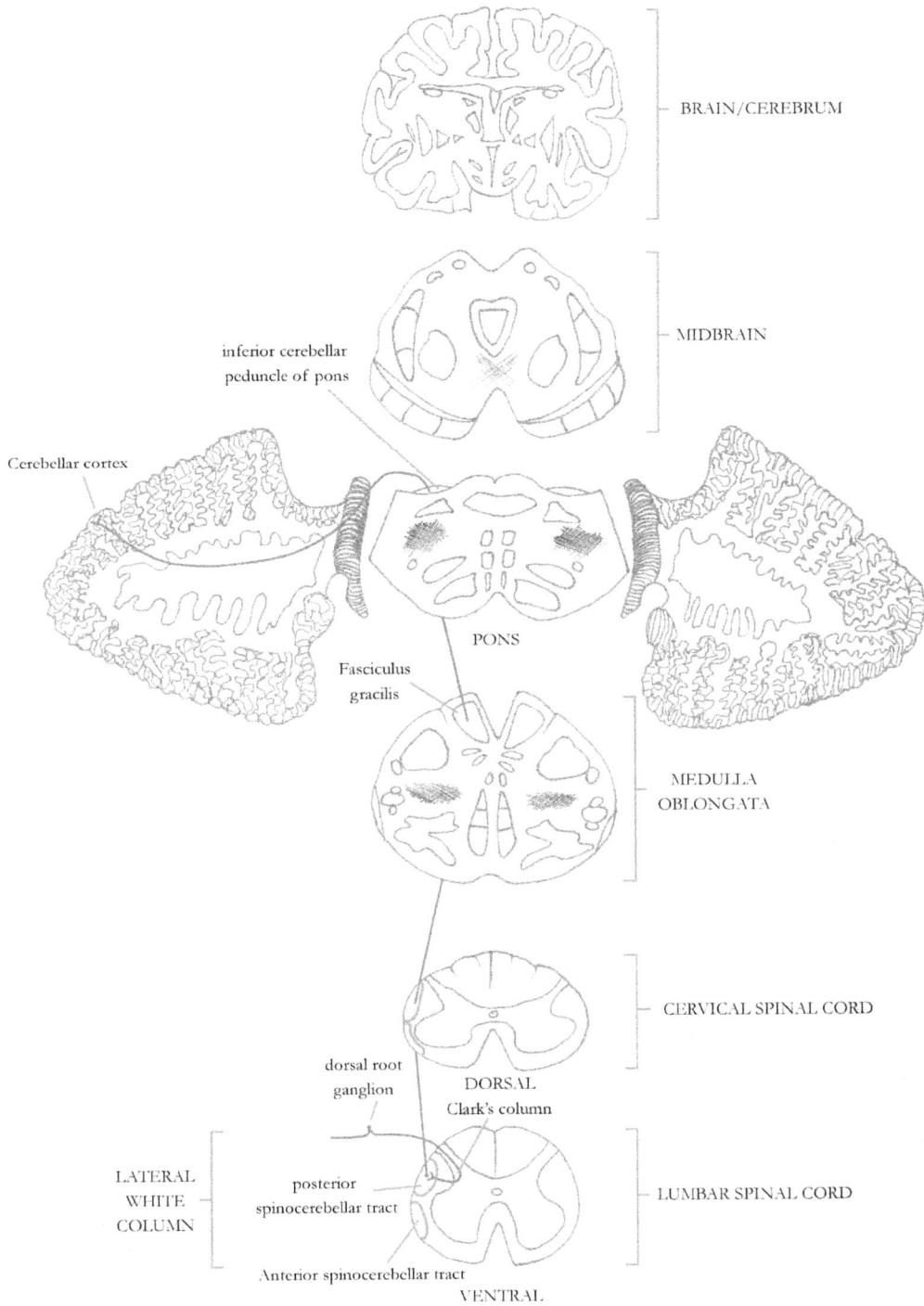

BRAIN/CEREBRUM

MIDBRAIN

inferior cerebellar
peduncle of pons

Cerebellar cortex

PONS

Fasciculus
gracilis

MEDULLA
OBLONGATA

CERVICAL SPINAL CORD

dorsal root
ganglion

DORSAL

Clark's column

LATERAL
WHITE
COLUMN

posterior
spinocerebellar tract

LUMBAR SPINAL CORD

Anterior spinocerebellar tract

VENTRAL

Anterior Spinocerebellar Tract

*Muscle/joint sense pathway to the cerebellum –
unconscious proprioception for the trunk,
as well as upper & lower limbs*

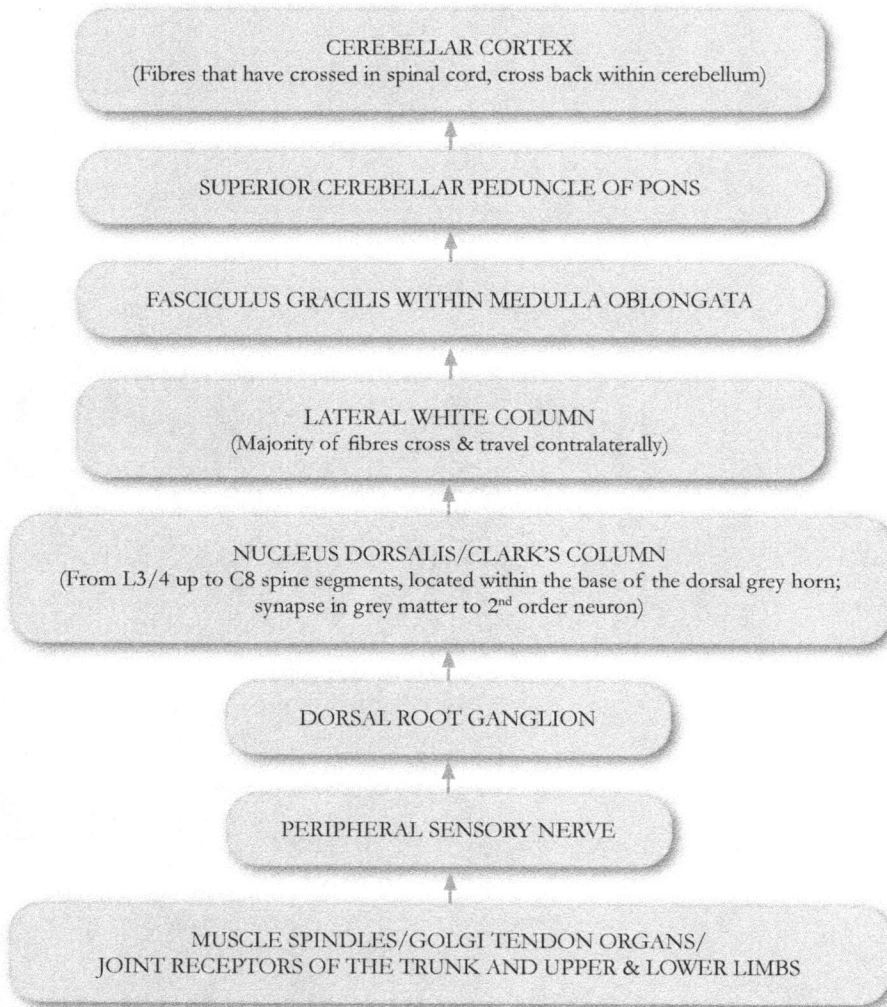

CEREBELLAR CORTEX
(Fibres that have crossed in spinal cord, cross back within cerebellum)

↑

SUPERIOR CEREBELLAR PEDUNCLE OF PONS

↑

FASCICULUS GRACILIS WITHIN MEDULLA OBLONGATA

↑

LATERAL WHITE COLUMN
(Majority of fibres cross & travel contralaterally)

↑

NUCLEUS DORSALIS/CLARK'S COLUMN
(From L3/4 up to C8 spine segments, located within the base of the dorsal grey horn;
synapse in grey matter to 2nd order neuron)

↑

DORSAL ROOT GANGLION

↑

PERIPHERAL SENSORY NERVE

↑

**MUSCLE SPINDLES/GOLGI TENDON ORGANS/
JOINT RECEPTORS OF THE TRUNK AND UPPER & LOWER LIMBS**

Ascending Sensory Tracts

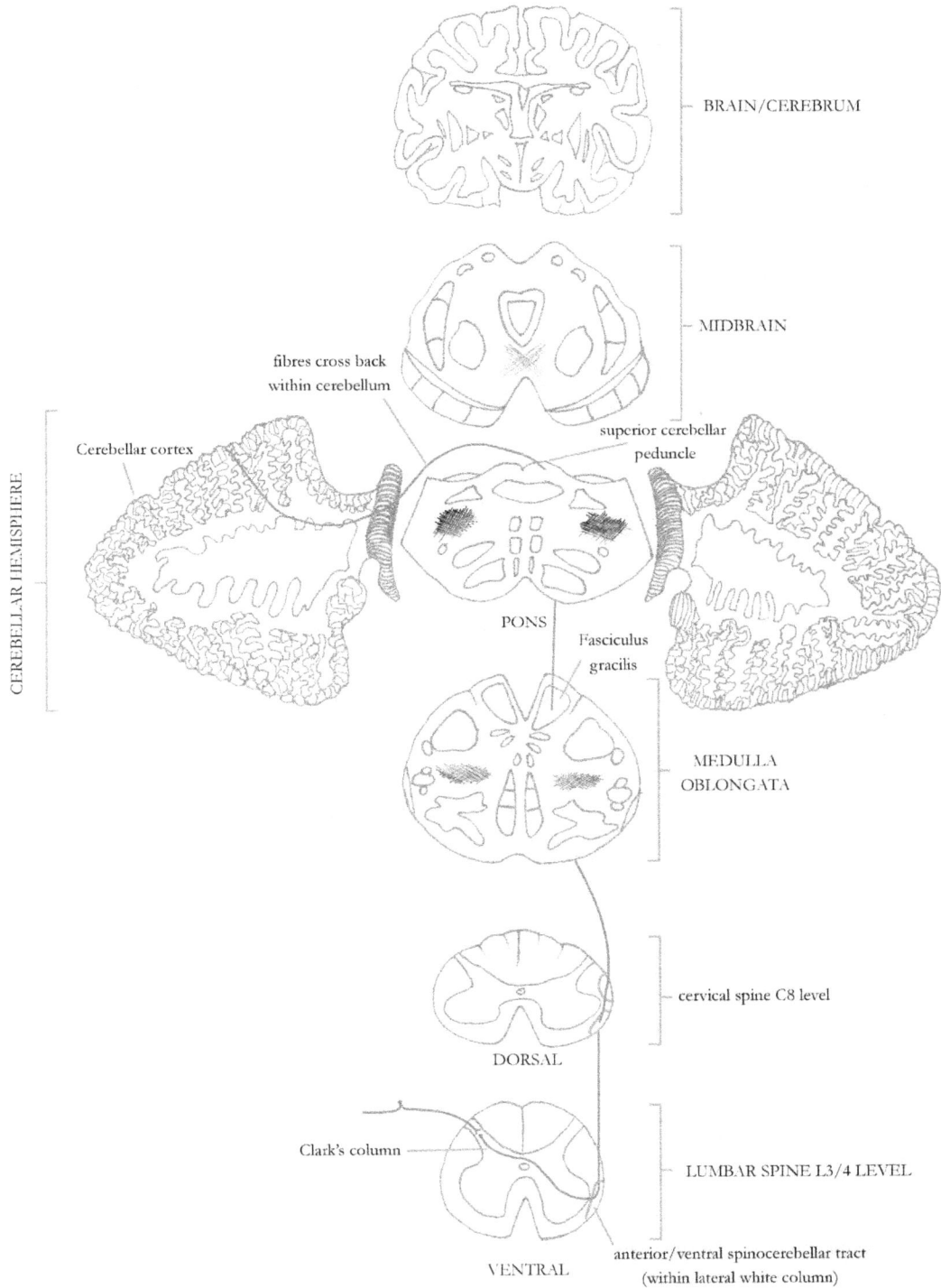

BRAIN/CEREBRUM

MIDBRAIN

fibres cross back
within cerebellum

superior cerebellar
peduncle

Cerebellar cortex

CEREBELLAR HEMISPHERE

PONS

Fasciculus
gracilis

MEDULLA
OBLONGATA

cervical spine C8 level

DORSAL

Clark's column

LUMBAR SPINE L3/4 LEVEL

anterior/ventral spinocerebellar tract
(within lateral white column)

VENTRAL

Cuneocerebellar Tract

Muscle/joint sense pathway to the cerebellum –
unconscious proprioception for upper trunk & upper limbs

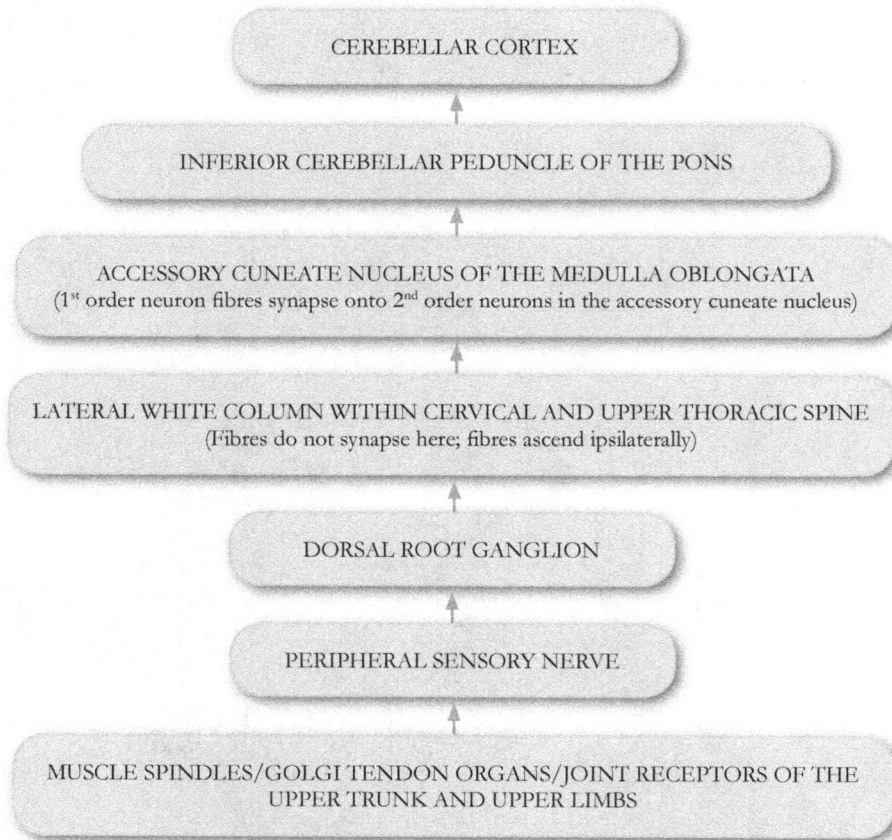

CEREBELLAR CORTEX

↑

INFERIOR CEREBELLAR PEDUNCLE OF THE PONS

↑

ACCESSORY CUNEATE NUCLEUS OF THE MEDULLA OBLONGATA
(1st order neuron fibres synapse onto 2nd order neurons in the accessory cuneate nucleus)

↑

LATERAL WHITE COLUMN WITHIN CERVICAL AND UPPER THORACIC SPINE
(Fibres do not synapse here; fibres ascend ipsilaterally)

↑

DORSAL ROOT GANGLION

↑

PERIPHERAL SENSORY NERVE

↑

MUSCLE SPINDLES/GOLGI TENDON ORGANS/JOINT RECEPTORS OF THE
UPPER TRUNK AND UPPER LIMBS

Ascending Sensory Tracts

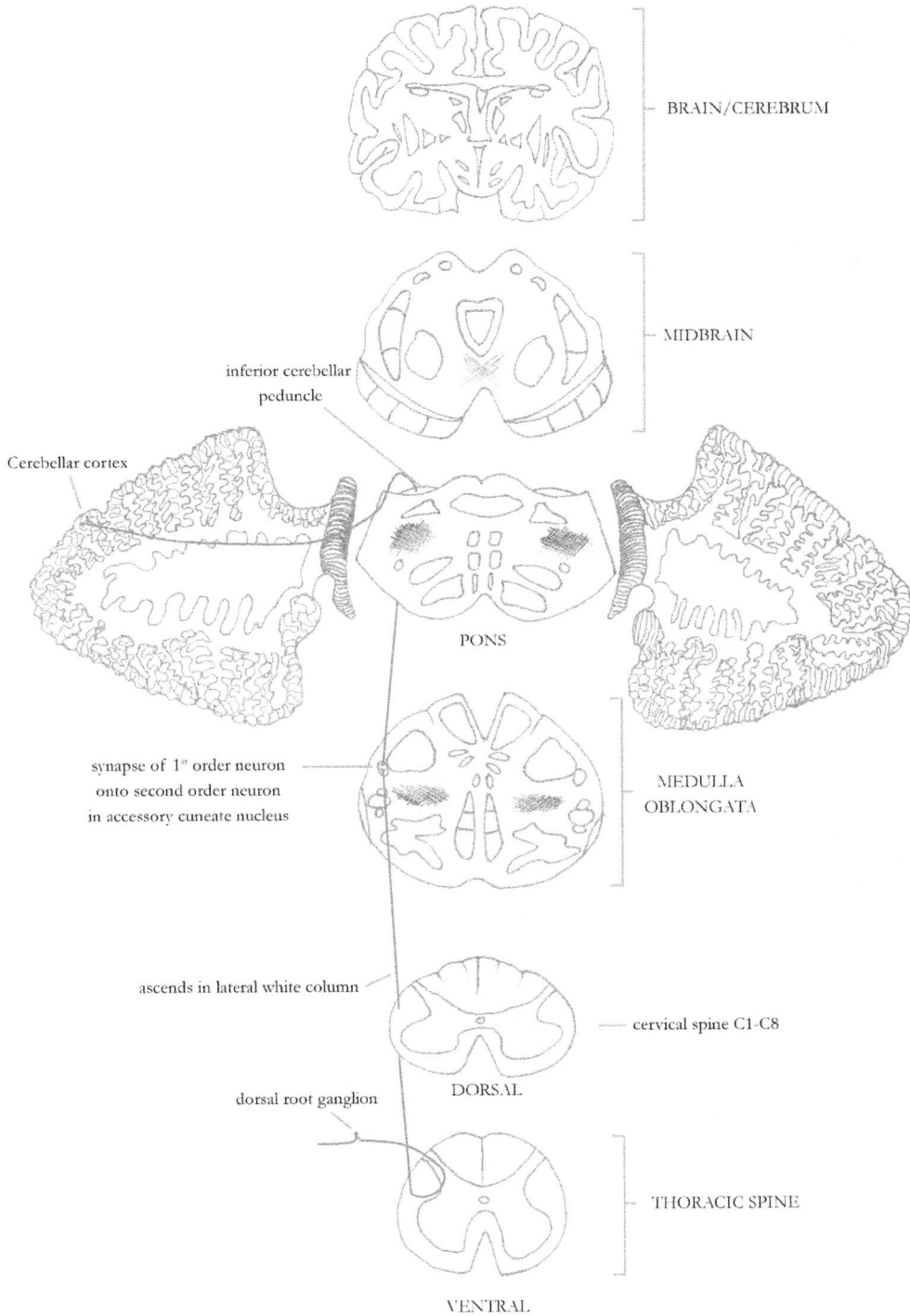

BRAIN/CEREBRUM

MIDBRAIN

inferior cerebellar
peduncle

Cerebellar cortex

PONS

synapse of 1ˢᵗ order neuron
onto second order neuron
in accessory cuneate nucleus

MEDULLA
OBLONGATA

ascends in lateral white column

cervical spine C1-C8

DORSAL

dorsal root ganglion

THORACIC SPINE

VENTRAL

Spinomesencephalic Pathways

Spinotectal tract, spinoannular tract & spinoparabrachial tract sensory information for spinovisual reflexes

SPINOANNULAR TRACT	SPINOTECTAL TRACT	SPINOPARABRACHIAL TRACT
↑	↑	↑
PERIAQUEDUCTAL GRAY WITHIN MIDBRAIN	SUPERIOR COLLICULUS WITHIN MIDBRAIN	PARABRACHIAL NUCLEI WITHIN SUPERIOR CEREBELLAR PEDUNCLE OF DORSOLATERAL PONS

(Synapse to 3rd order neuron)

↑

MEDULLA OBLONGATA & PONS

↑

FIBRES ASCEND AS SPINOTECTAL TRACT IN ANTEROLATERAL WHITE COLUMN

↑

GREY MATTER OF SPINAL CORD WITHIN LAMINA I, IV-VIII, CONCENTRATED IN LAMINA V
(Synapse to 2nd order neuron)

↑

DORSAL ROOT GANGLION

↑

PERIPHERAL SENSORY NERVE

↑

VARIOUS SENSORY RECEPTORS IN THE PERIPHERY

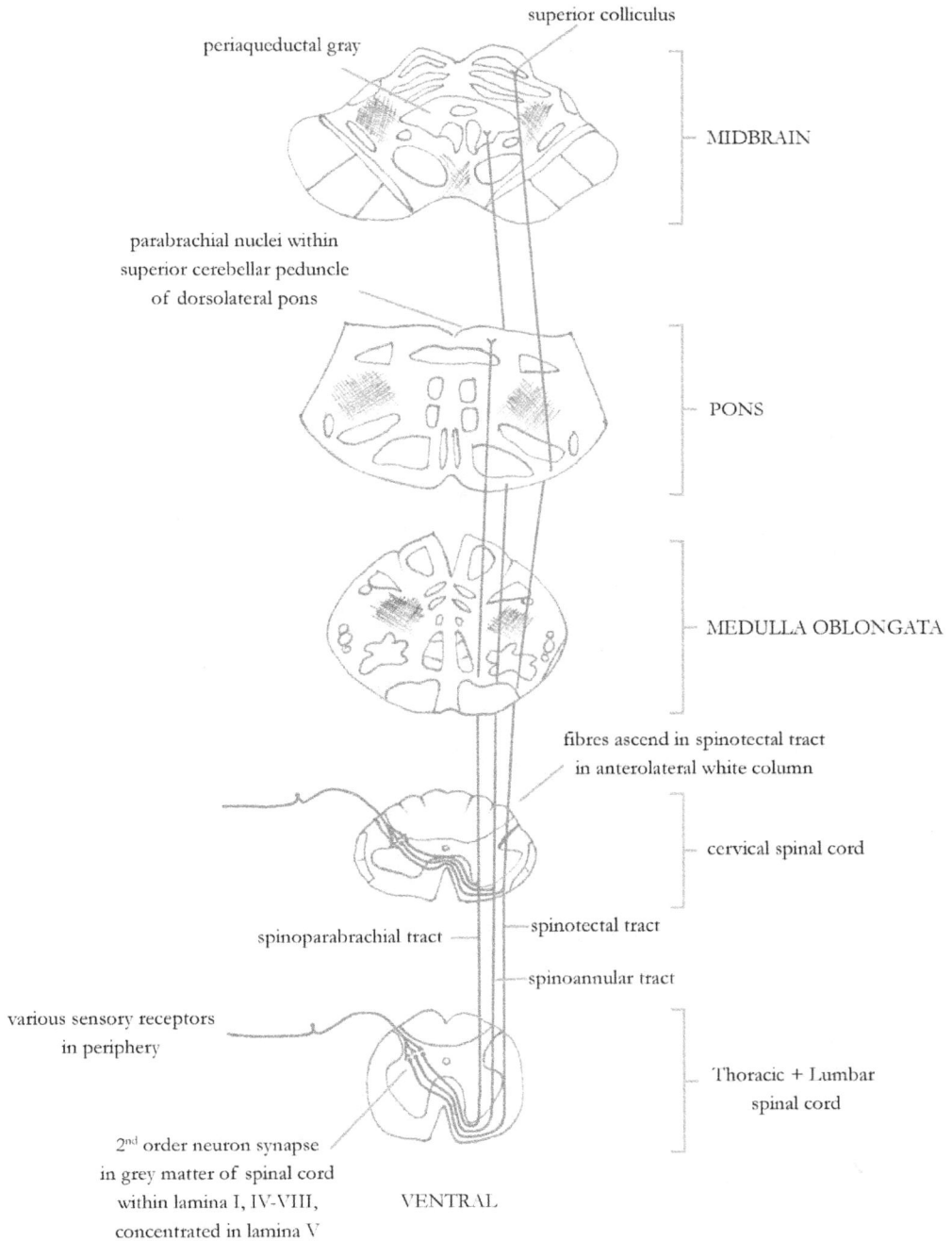

superior colliculus

periaqueductal gray

MIDBRAIN

parabrachial nuclei within
superior cerebellar peduncle
of dorsolateral pons

PONS

MEDULLA OBLONGATA

fibres ascend in spinotectal tract
in anterolateral white column

cervical spinal cord

spinotectal tract

spinoparabrachial tract

spinoannular tract

various sensory receptors
in periphery

Thoracic + Lumbar
spinal cord

2nd order neuron synapse
in grey matter of spinal cord
within lamina I, IV-VIII,
concentrated in lamina V

VENTRAL

Spinoreticular Tract

Sensory pathway to reticular formation/conscious perception

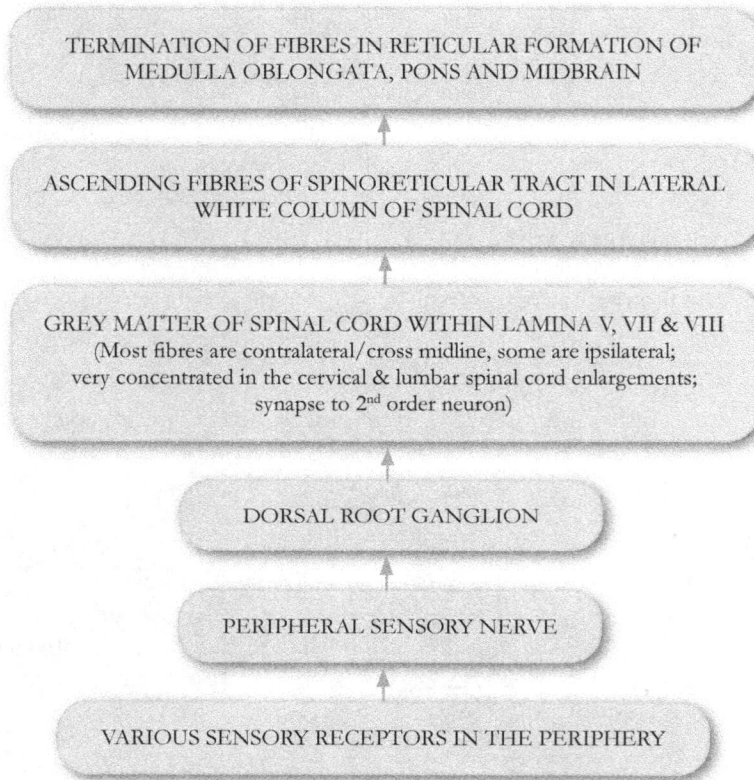

TERMINATION OF FIBRES IN RETICULAR FORMATION OF
MEDULLA OBLONGATA, PONS AND MIDBRAIN

↑

ASCENDING FIBRES OF SPINORETICULAR TRACT IN LATERAL
WHITE COLUMN OF SPINAL CORD

↑

GREY MATTER OF SPINAL CORD WITHIN LAMINA V, VII & VIII
(Most fibres are contralateral/cross midline, some are ipsilateral;
very concentrated in the cervical & lumbar spinal cord enlargements;
synapse to 2nd order neuron)

↑

DORSAL ROOT GANGLION

↑

PERIPHERAL SENSORY NERVE

↑

VARIOUS SENSORY RECEPTORS IN THE PERIPHERY

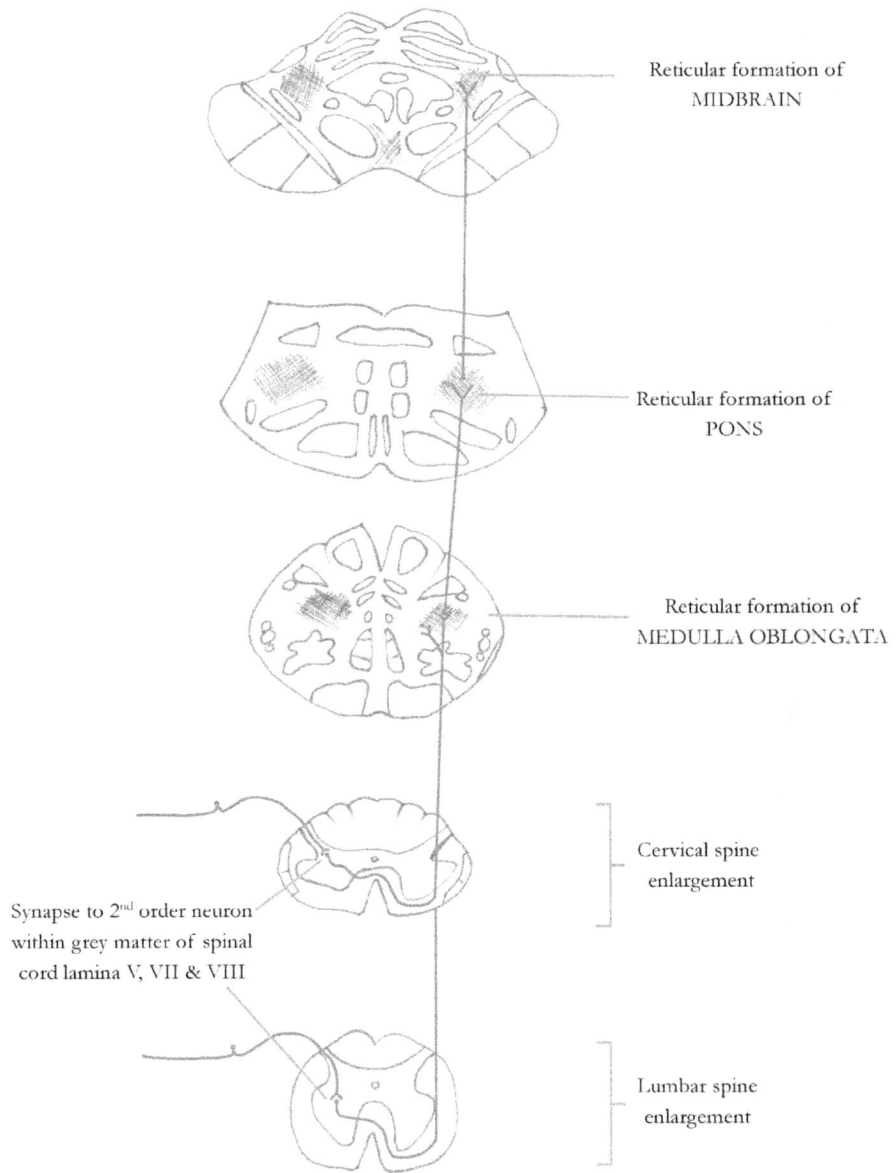

Reticular formation of
MIDBRAIN

Reticular formation of
PONS

Reticular formation of
MEDULLA OBLONGATA

Cervical spine
enlargement

Synapse to 2nd order neuron
within grey matter of spinal
cord lamina V, VII & VIII

Lumbar spine
enlargement

Spino-Olivary Tract (Helweg's Tract)

Proprioception of all limbs

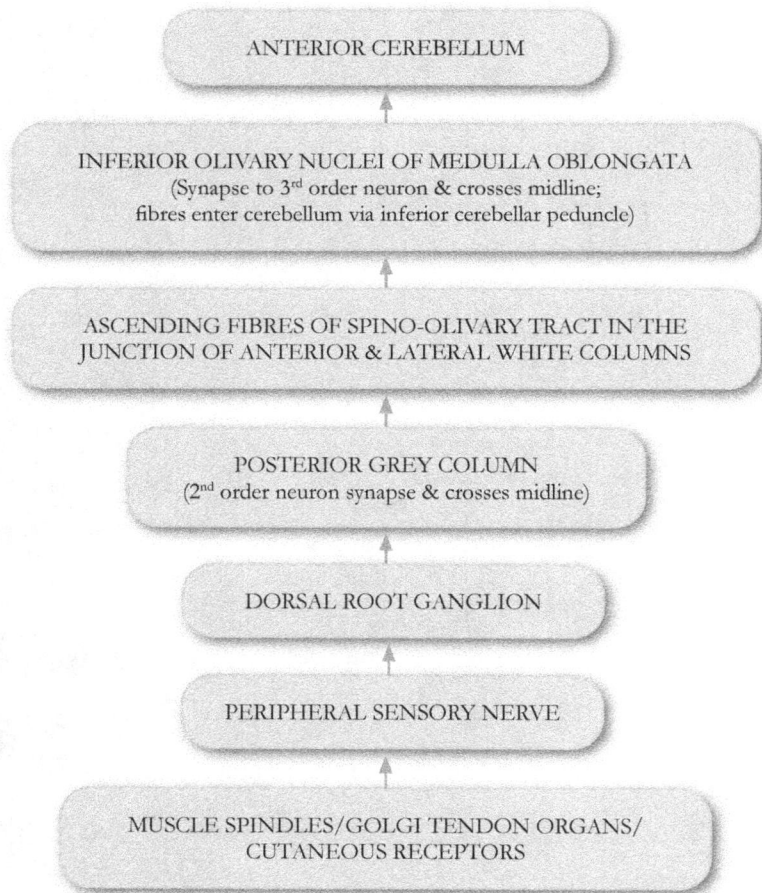

ANTERIOR CEREBELLUM

↑

INFERIOR OLIVARY NUCLEI OF MEDULLA OBLONGATA
(Synapse to 3rd order neuron & crosses midline;
fibres enter cerebellum via inferior cerebellar peduncle)

↑

ASCENDING FIBRES OF SPINO-OLIVARY TRACT IN THE
JUNCTION OF ANTERIOR & LATERAL WHITE COLUMNS

↑

POSTERIOR GREY COLUMN
(2nd order neuron synapse & crosses midline)

↑

DORSAL ROOT GANGLION

↑

PERIPHERAL SENSORY NERVE

↑

MUSCLE SPINDLES/GOLGI TENDON ORGANS/
CUTANEOUS RECEPTORS

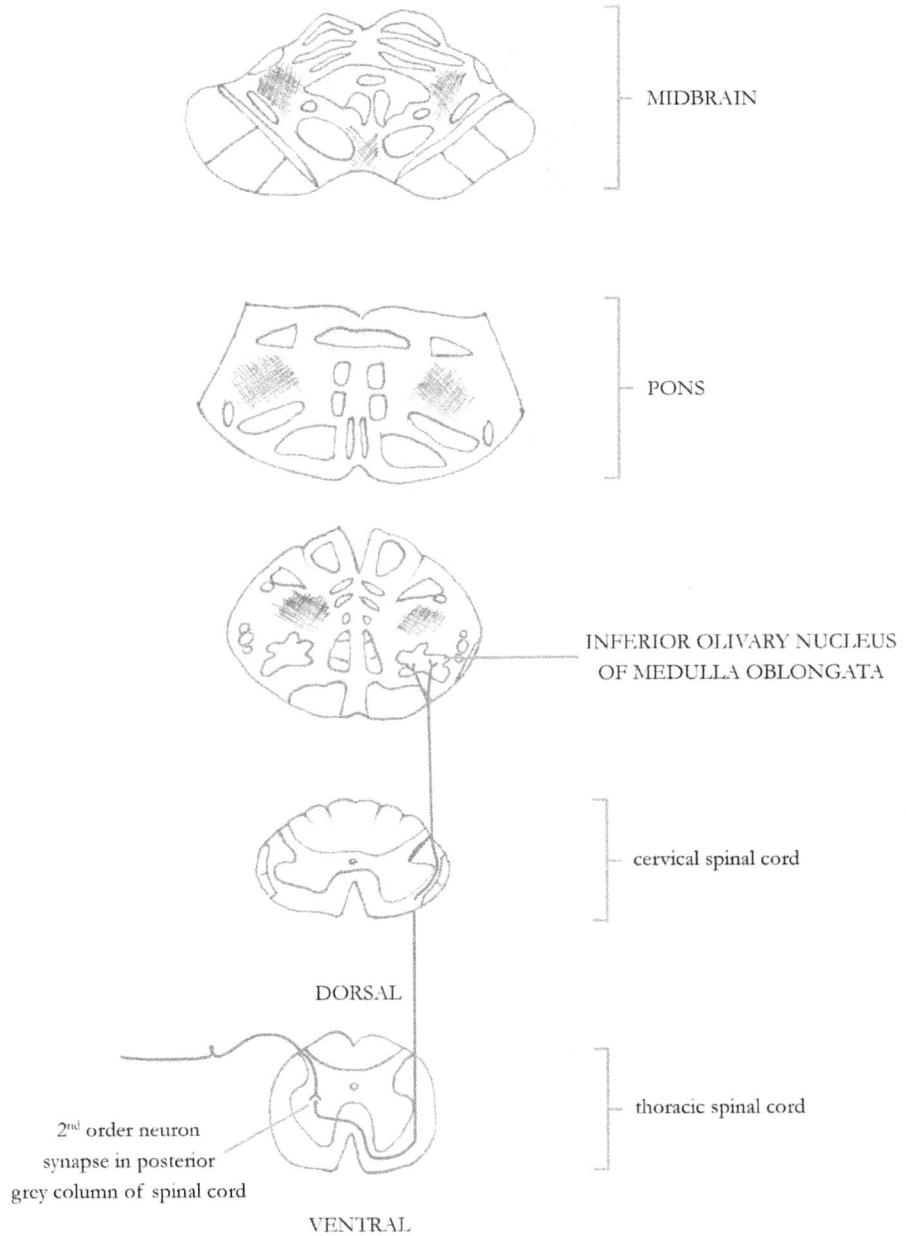

MIDBRAIN

PONS

INFERIOR OLIVARY NUCLEUS
OF MEDULLA OBLONGATA

cervical spinal cord

DORSAL

thoracic spinal cord

2nd order neuron
synapse in posterior
grey column of spinal cord

VENTRAL

Dorsal Spino-Olivocerebellar Tract

Cutaneous proprioception of distal limbs
Proprioception of all limbs

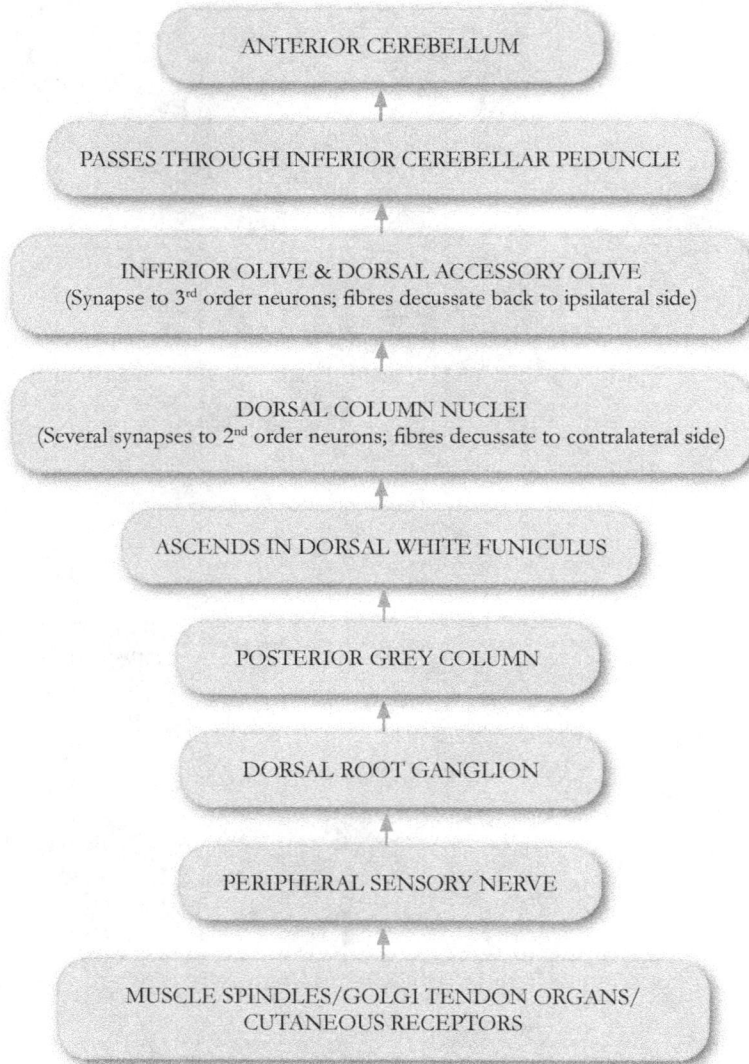

ANTERIOR CEREBELLUM

↑

PASSES THROUGH INFERIOR CEREBELLAR PEDUNCLE

↑

INFERIOR OLIVE & DORSAL ACCESSORY OLIVE
(Synapse to 3rd order neurons; fibres decussate back to ipsilateral side)

↑

DORSAL COLUMN NUCLEI
(Several synapses to 2nd order neurons; fibres decussate to contralateral side)

↑

ASCENDS IN DORSAL WHITE FUNICULUS

↑

POSTERIOR GREY COLUMN

↑

DORSAL ROOT GANGLION

↑

PERIPHERAL SENSORY NERVE

↑

MUSCLE SPINDLES/GOLGI TENDON ORGANS/
CUTANEOUS RECEPTORS

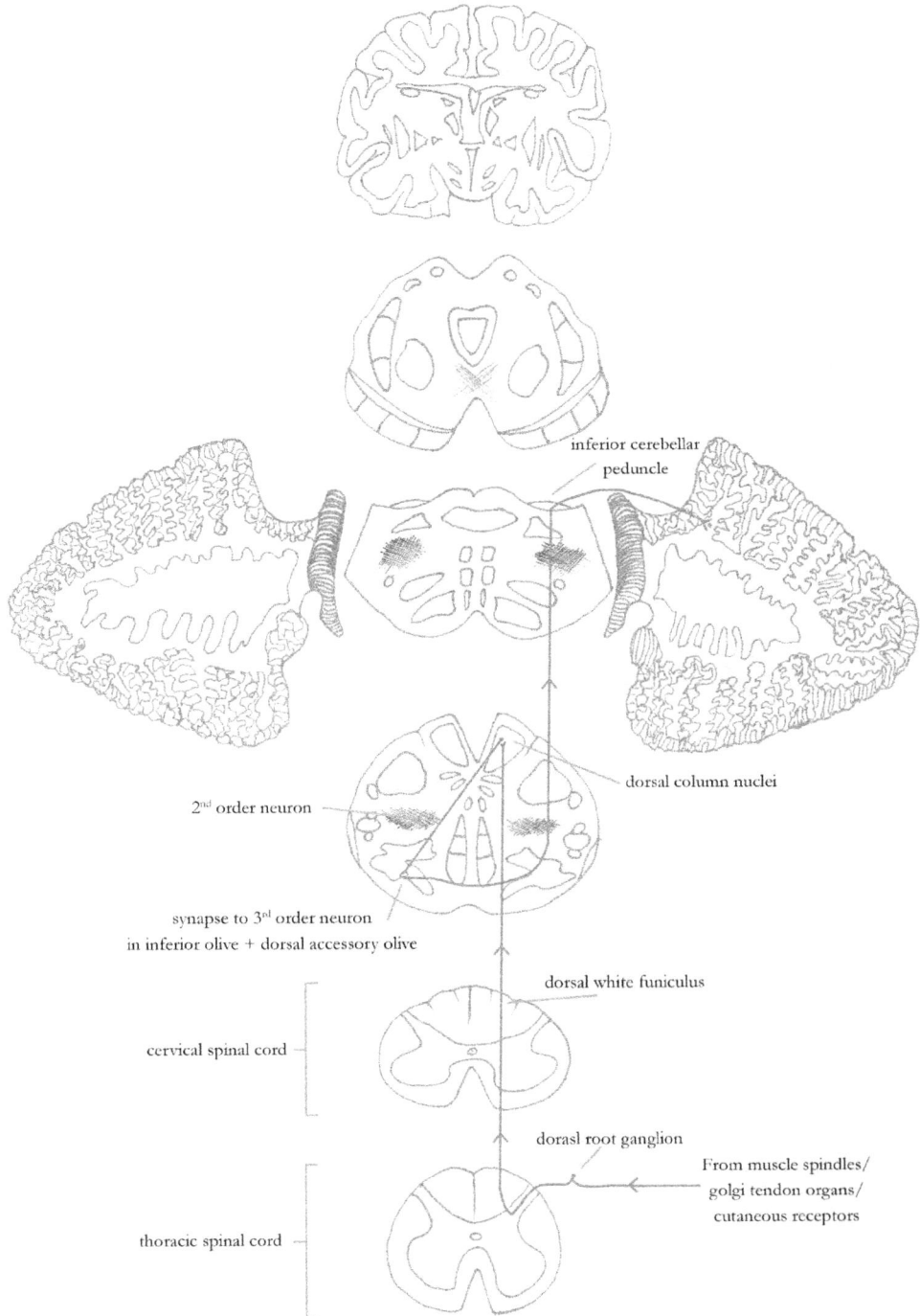

inferior cerebellar
peduncle

dorsal column nuclei

2nd order neuron

synapse to 3rd order neuron
in inferior olive + dorsal accessory olive

dorsal white funiculus

cervical spinal cord

dorsal root ganglion

From muscle spindles/
golgi tendon organs/
cutaneous receptors

thoracic spinal cord

DESCENDING MOTOR TRACTS

DESCENDING MOTOR TRACTS
(Pyramidal and extrapyramidal pathways)

PYRAMIDAL TRACT (Pass through the pyramids of medulla oblongata)
1. Corticospinal

EXTRAPYRAMIDAL TRACTS (Pass though areas outside of the pyramids of the medulla oblongata)
1. Reticulospinal Tract
2. Tectospinal Tract
3. Rubrospinal Tract
4. Vestibulospinal Tract
5. Olivospinal Tract
6. Interstitiospinal Tract
7. Solitariospinal Tract
8. Descending Autonomic Tracts

Corticospinal Tract

Sole pathway of voluntary movement

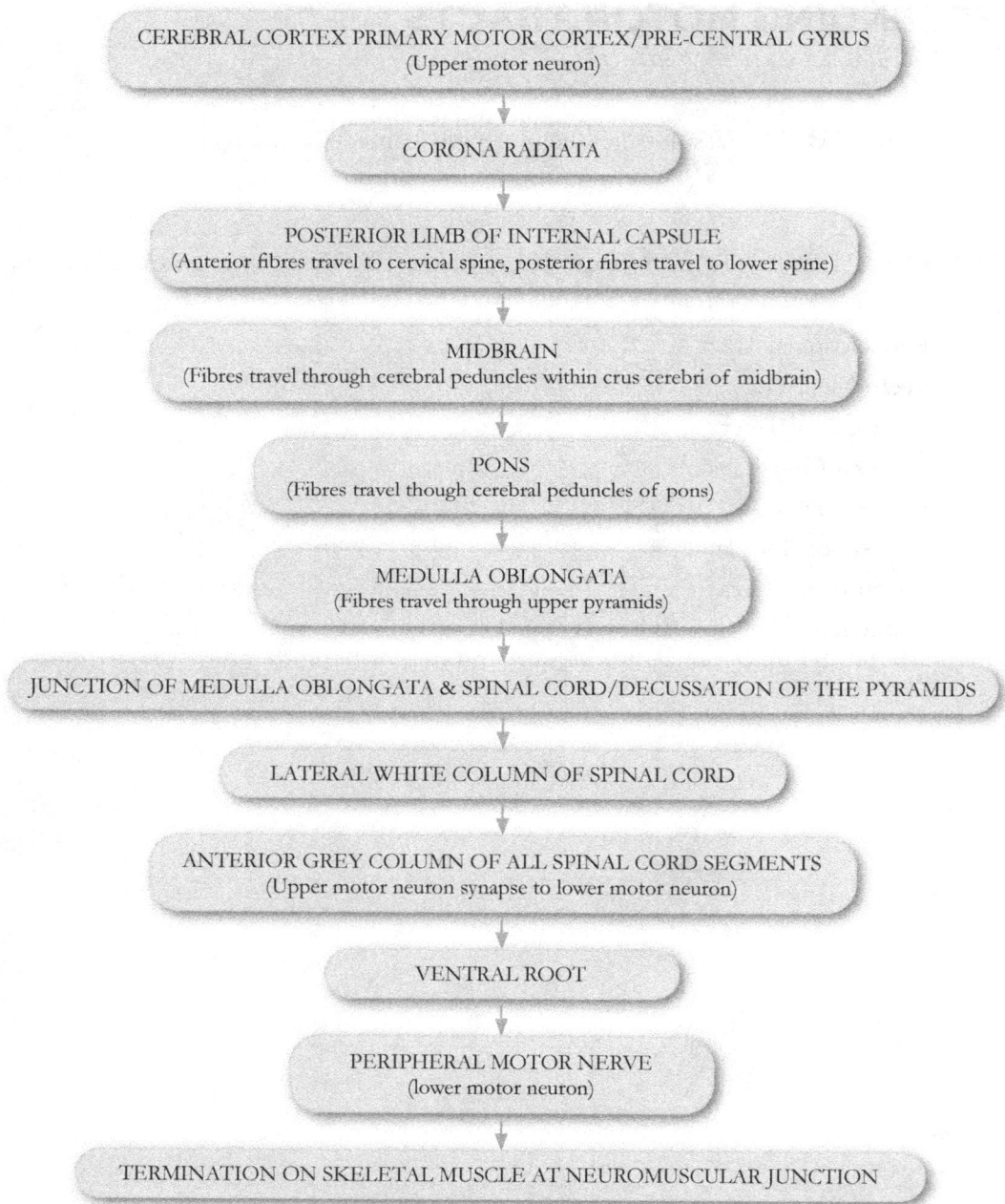

CEREBRAL CORTEX PRIMARY MOTOR CORTEX/PRE-CENTRAL GYRUS
(Upper motor neuron)

CORONA RADIATA

POSTERIOR LIMB OF INTERNAL CAPSULE
(Anterior fibres travel to cervical spine, posterior fibres travel to lower spine)

MIDBRAIN
(Fibres travel through cerebral peduncles within crus cerebri of midbrain)

PONS
(Fibres travel though cerebral peduncles of pons)

MEDULLA OBLONGATA
(Fibres travel through upper pyramids)

JUNCTION OF MEDULLA OBLONGATA & SPINAL CORD/DECUSSATION OF THE PYRAMIDS

LATERAL WHITE COLUMN OF SPINAL CORD

ANTERIOR GREY COLUMN OF ALL SPINAL CORD SEGMENTS
(Upper motor neuron synapse to lower motor neuron)

VENTRAL ROOT

PERIPHERAL MOTOR NERVE
(lower motor neuron)

TERMINATION ON SKELETAL MUSCLE AT NEUROMUSCULAR JUNCTION

Descending Motor Tracts

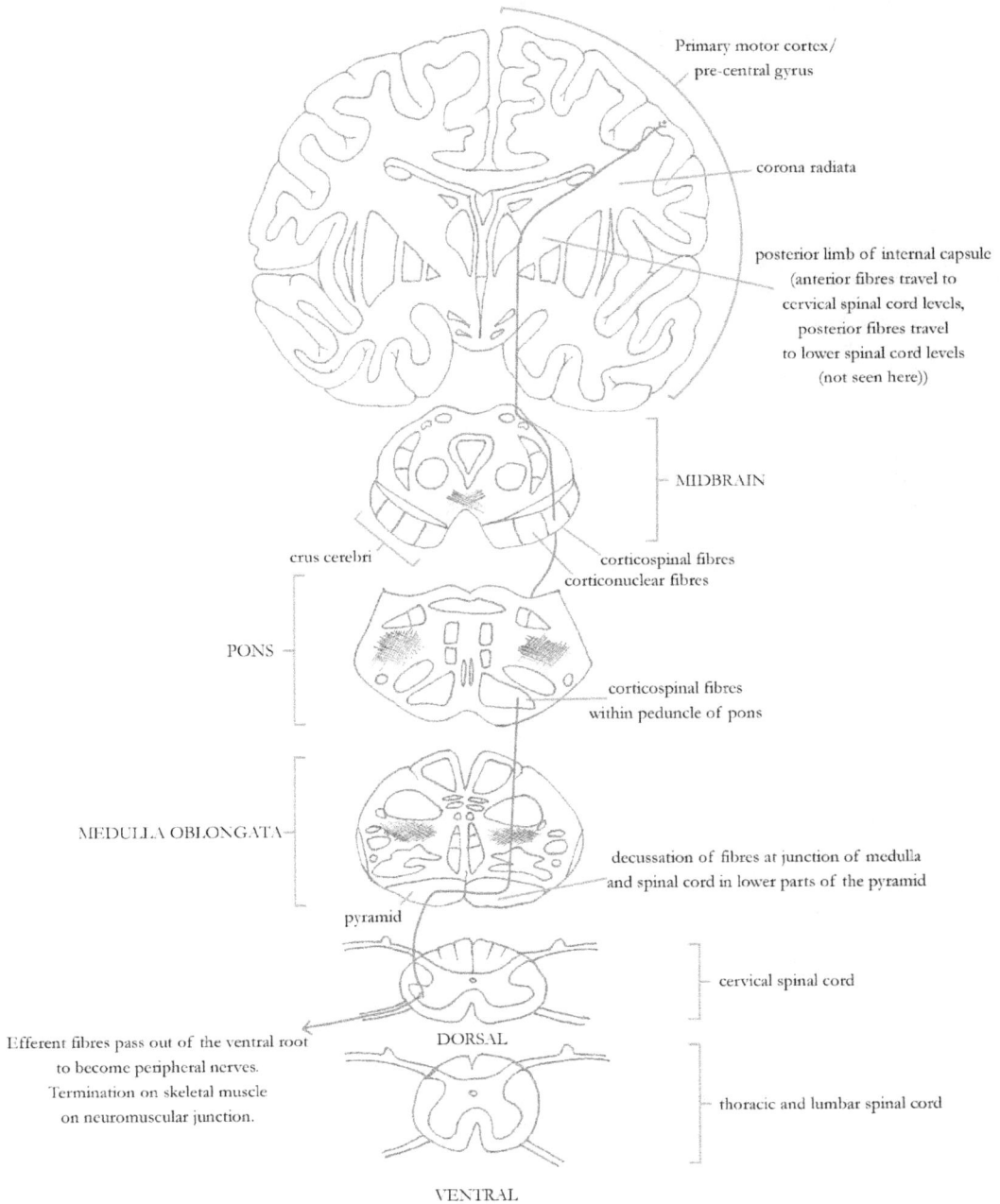

Primary motor cortex/
pre-central gyrus

corona radiata

posterior limb of internal capsule
(anterior fibres travel to
cervical spinal cord levels,
posterior fibres travel
to lower spinal cord levels
(not seen here))

MIDBRAIN

crus cerebri

corticospinal fibres
corticonuclear fibres

PONS

corticospinal fibres
within peduncle of pons

MEDULLA OBLONGATA

decussation of fibres at junction of medulla
and spinal cord in lower parts of the pyramid

pyramid

cervical spinal cord

Efferent fibres pass out of the ventral root
to become peripheral nerves.
Termination on skeletal muscle
on neuromuscular junction.

DORSAL

thoracic and lumbar spinal cord

VENTRAL

Medial/Pontine Reticulospinal Tract
and
Lateral/Medullary/Bulbo Reticulospinal Tract
With Cortical Projections

Sole pathway of voluntary movement

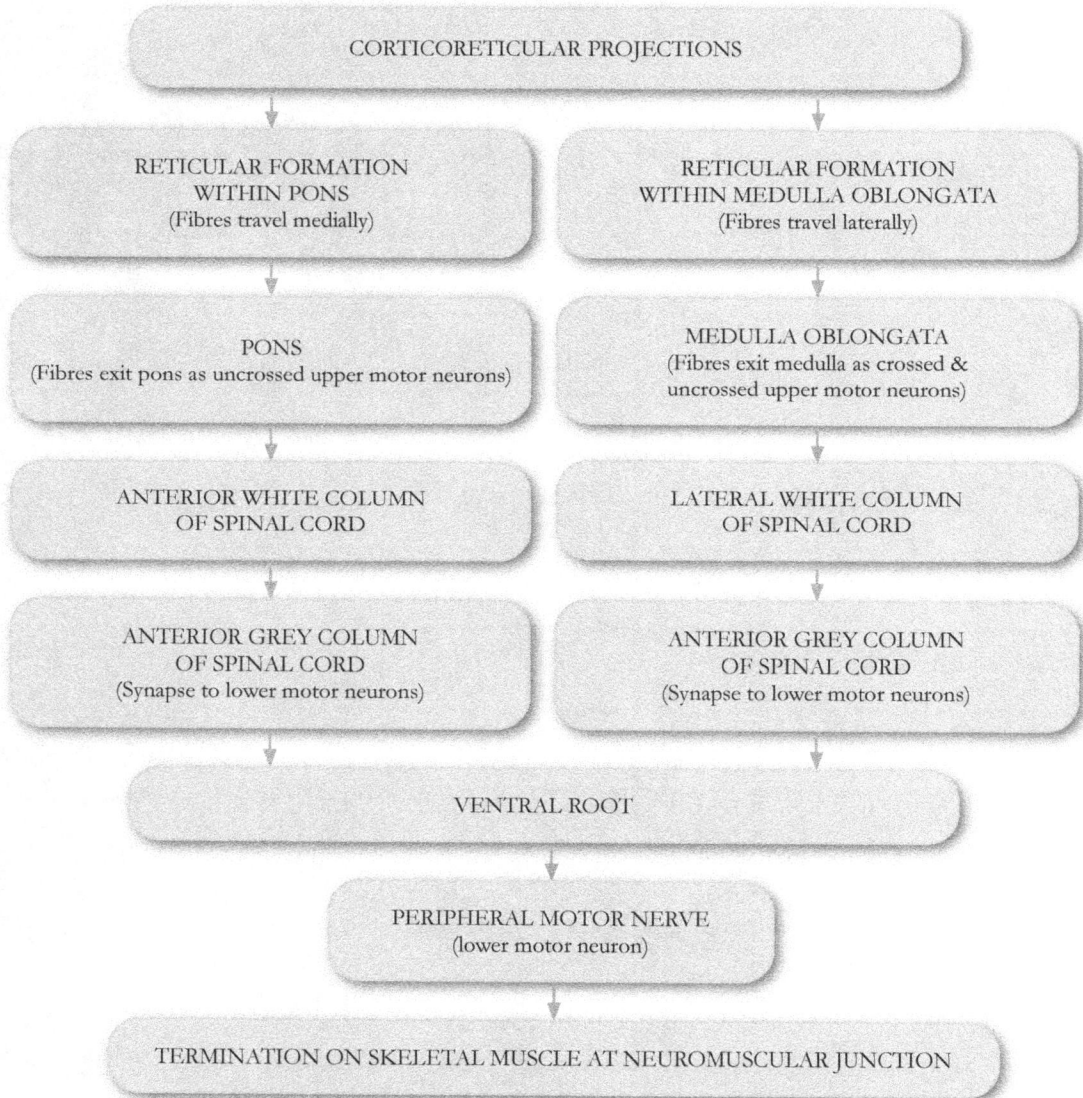

CORTICORETICULAR PROJECTIONS

RETICULAR FORMATION
WITHIN PONS
(Fibres travel medially)

RETICULAR FORMATION
WITHIN MEDULLA OBLONGATA
(Fibres travel laterally)

PONS
(Fibres exit pons as uncrossed upper motor neurons)

MEDULLA OBLONGATA
(Fibres exit medulla as crossed &
uncrossed upper motor neurons)

ANTERIOR WHITE COLUMN
OF SPINAL CORD

LATERAL WHITE COLUMN
OF SPINAL CORD

ANTERIOR GREY COLUMN
OF SPINAL CORD
(Synapse to lower motor neurons)

ANTERIOR GREY COLUMN
OF SPINAL CORD
(Synapse to lower motor neurons)

VENTRAL ROOT

PERIPHERAL MOTOR NERVE
(lower motor neuron)

TERMINATION ON SKELETAL MUSCLE AT NEUROMUSCULAR JUNCTION

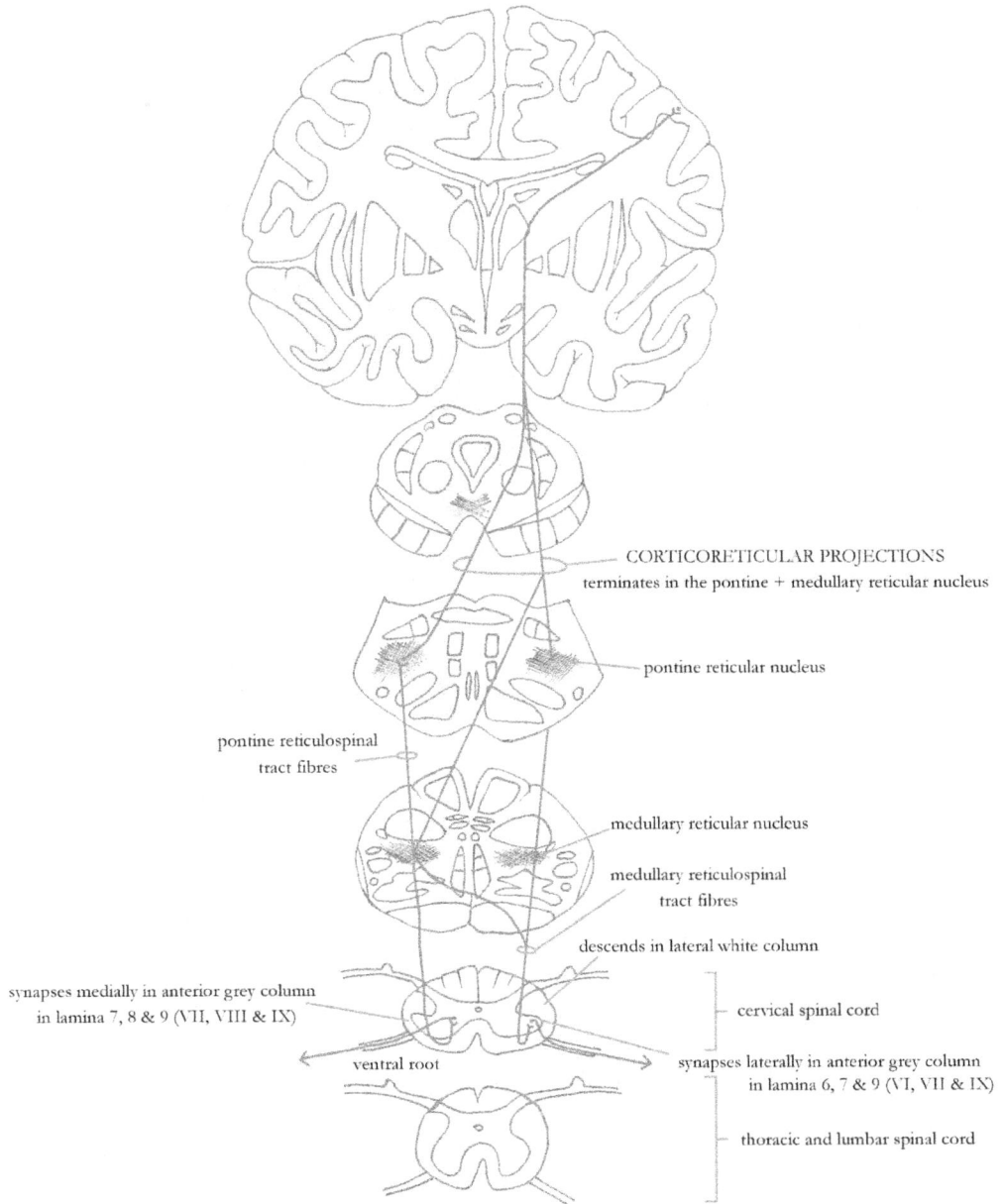

CORTICORETICULAR PROJECTIONS
terminates in the pontine + medullary reticular nucleus

pontine reticular nucleus

pontine reticulospinal
tract fibres

medullary reticular nucleus

medullary reticulospinal
tract fibres

descends in lateral white column

synapses medially in anterior grey column
in lamina 7, 8 & 9 (VII, VIII & IX)

cervical spinal cord

ventral root

synapses laterally in anterior grey column
in lamina 6, 7 & 9 (VI, VII & IX)

thoracic and lumbar spinal cord

Tectospinal Tract
With Cortical Projections

Reflex postural movements of the head in response to visual stimuli

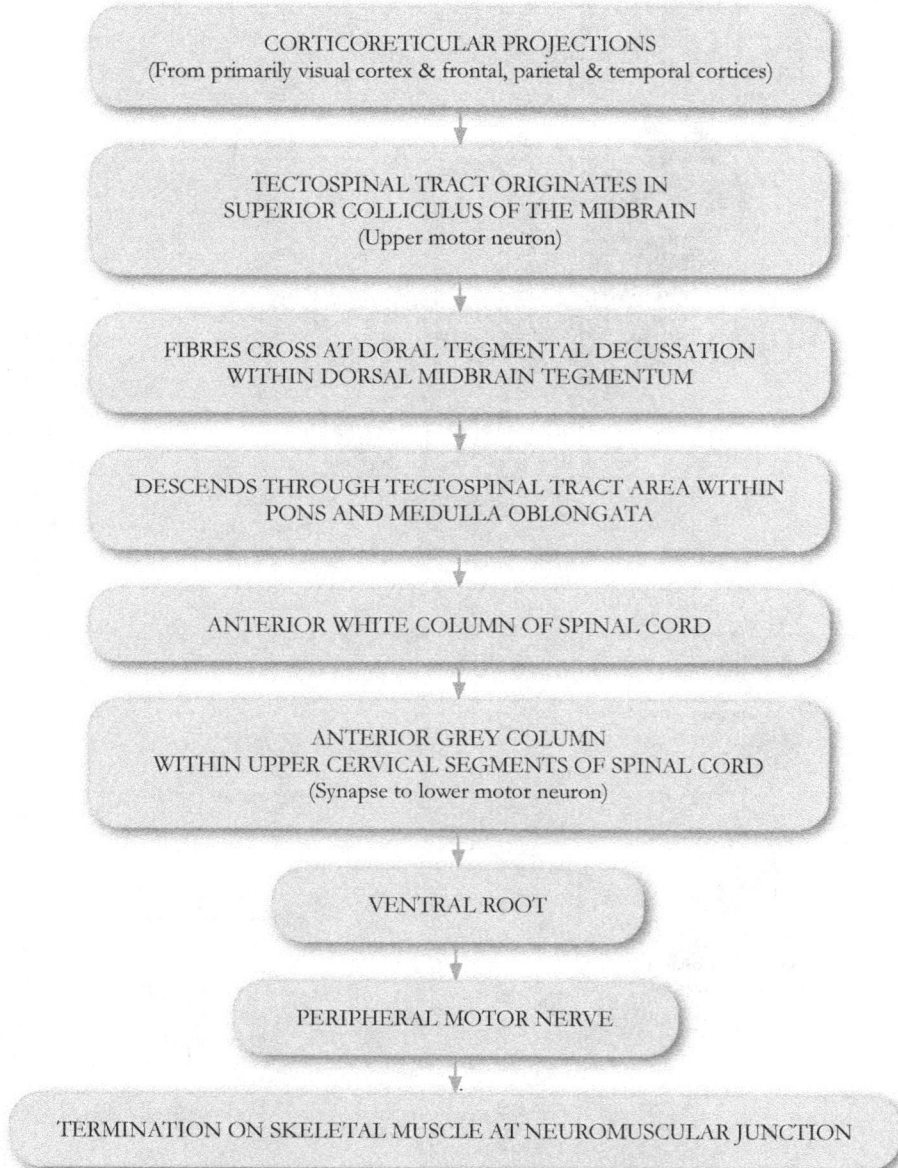

CORTICORETICULAR PROJECTIONS
(From primarily visual cortex & frontal, parietal & temporal cortices)

↓

TECTOSPINAL TRACT ORIGINATES IN
SUPERIOR COLLICULUS OF THE MIDBRAIN
(Upper motor neuron)

↓

FIBRES CROSS AT DORAL TEGMENTAL DECUSSATION
WITHIN DORSAL MIDBRAIN TEGMENTUM

↓

DESCENDS THROUGH TECTOSPINAL TRACT AREA WITHIN
PONS AND MEDULLA OBLONGATA

↓

ANTERIOR WHITE COLUMN OF SPINAL CORD

↓

ANTERIOR GREY COLUMN
WITHIN UPPER CERVICAL SEGMENTS OF SPINAL CORD
(Synapse to lower motor neuron)

↓

VENTRAL ROOT

↓

PERIPHERAL MOTOR NERVE

↓

TERMINATION ON SKELETAL MUSCLE AT NEUROMUSCULAR JUNCTION

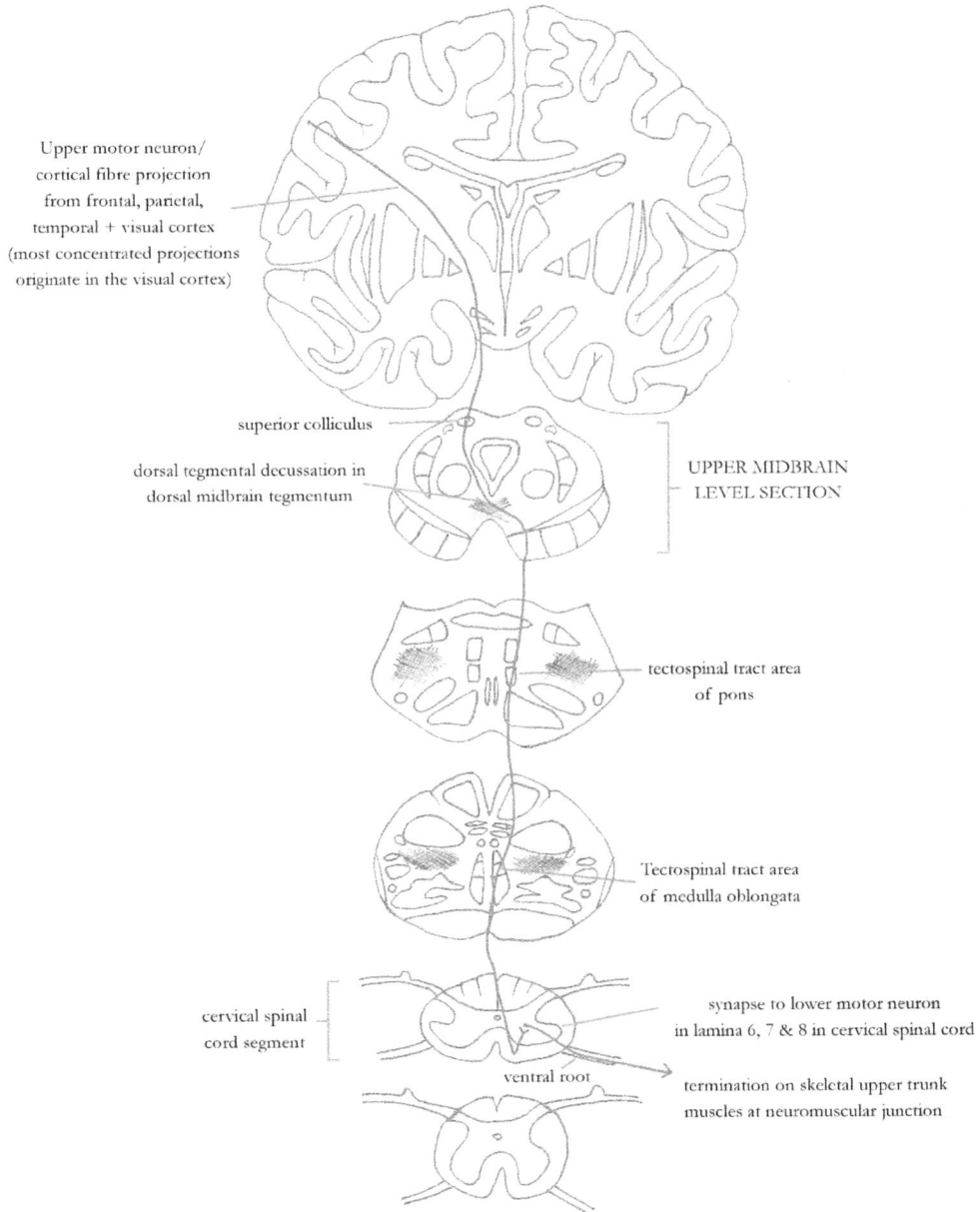

Upper motor neuron/
cortical fibre projection
from frontal, parietal,
temporal + visual cortex
(most concentrated projections
originate in the visual cortex)

superior colliculus

dorsal tegmental decussation in
dorsal midbrain tegmentum

UPPER MIDBRAIN
LEVEL SECTION

tectospinal tract area
of pons

Tectospinal tract area
of medulla oblongata

cervical spinal
cord segment

synapse to lower motor neuron
in lamina 6, 7 & 8 in cervical spinal cord

ventral root

termination on skeletal upper trunk
muscles at neuromuscular junction

Tectobulbar Tract

Coordination of head and eye movements

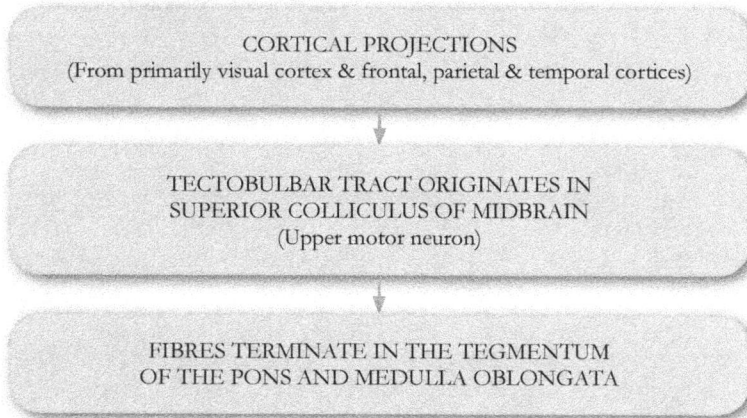

CORTICAL PROJECTIONS
(From primarily visual cortex & frontal, parietal & temporal cortices)

TECTOBULBAR TRACT ORIGINATES IN
SUPERIOR COLLICULUS OF MIDBRAIN
(Upper motor neuron)

FIBRES TERMINATE IN THE TEGMENTUM
OF THE PONS AND MEDULLA OBLONGATA

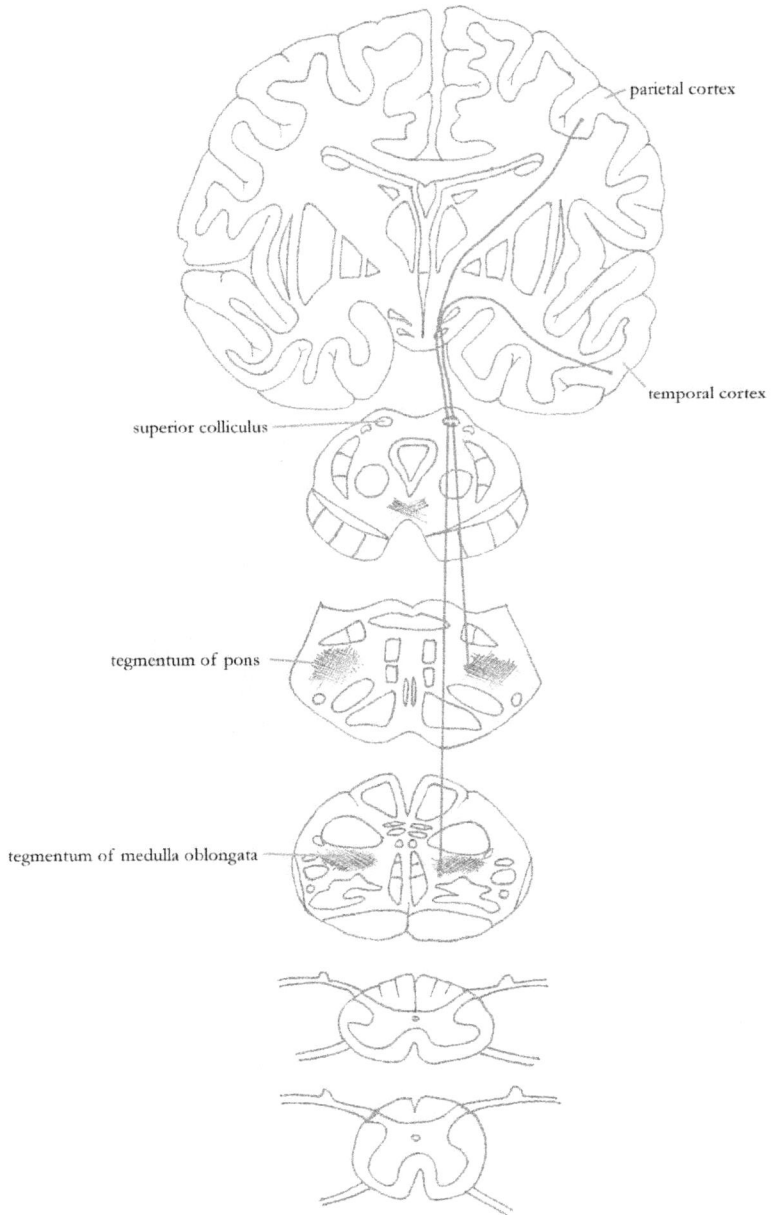

parietal cortex

temporal cortex

superior colliculus

tegmentum of pons

tegmentum of medulla oblongata

Rubrospinal, Rubroreticular & Rubroolivary Tracts

Facilitates activity of flexor muscles and inhibits extensor muscles

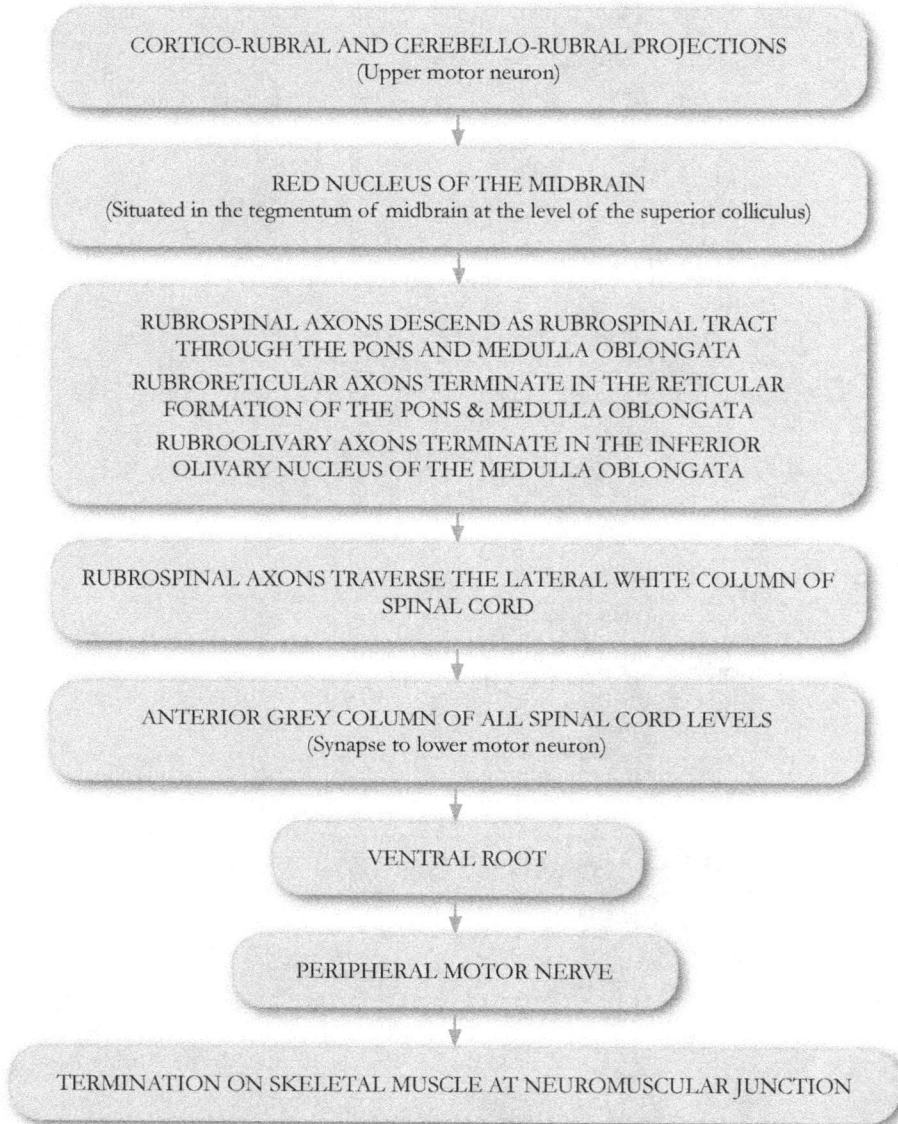

CORTICO-RUBRAL AND CEREBELLO-RUBRAL PROJECTIONS
(Upper motor neuron)

RED NUCLEUS OF THE MIDBRAIN
(Situated in the tegmentum of midbrain at the level of the superior colliculus)

RUBROSPINAL AXONS DESCEND AS RUBROSPINAL TRACT
THROUGH THE PONS AND MEDULLA OBLONGATA
RUBRORETICULAR AXONS TERMINATE IN THE RETICULAR
FORMATION OF THE PONS & MEDULLA OBLONGATA
RUBROOLIVARY AXONS TERMINATE IN THE INFERIOR
OLIVARY NUCLEUS OF THE MEDULLA OBLONGATA

RUBROSPINAL AXONS TRAVERSE THE LATERAL WHITE COLUMN OF
SPINAL CORD

ANTERIOR GREY COLUMN OF ALL SPINAL CORD LEVELS
(Synapse to lower motor neuron)

VENTRAL ROOT

PERIPHERAL MOTOR NERVE

TERMINATION ON SKELETAL MUSCLE AT NEUROMUSCULAR JUNCTION

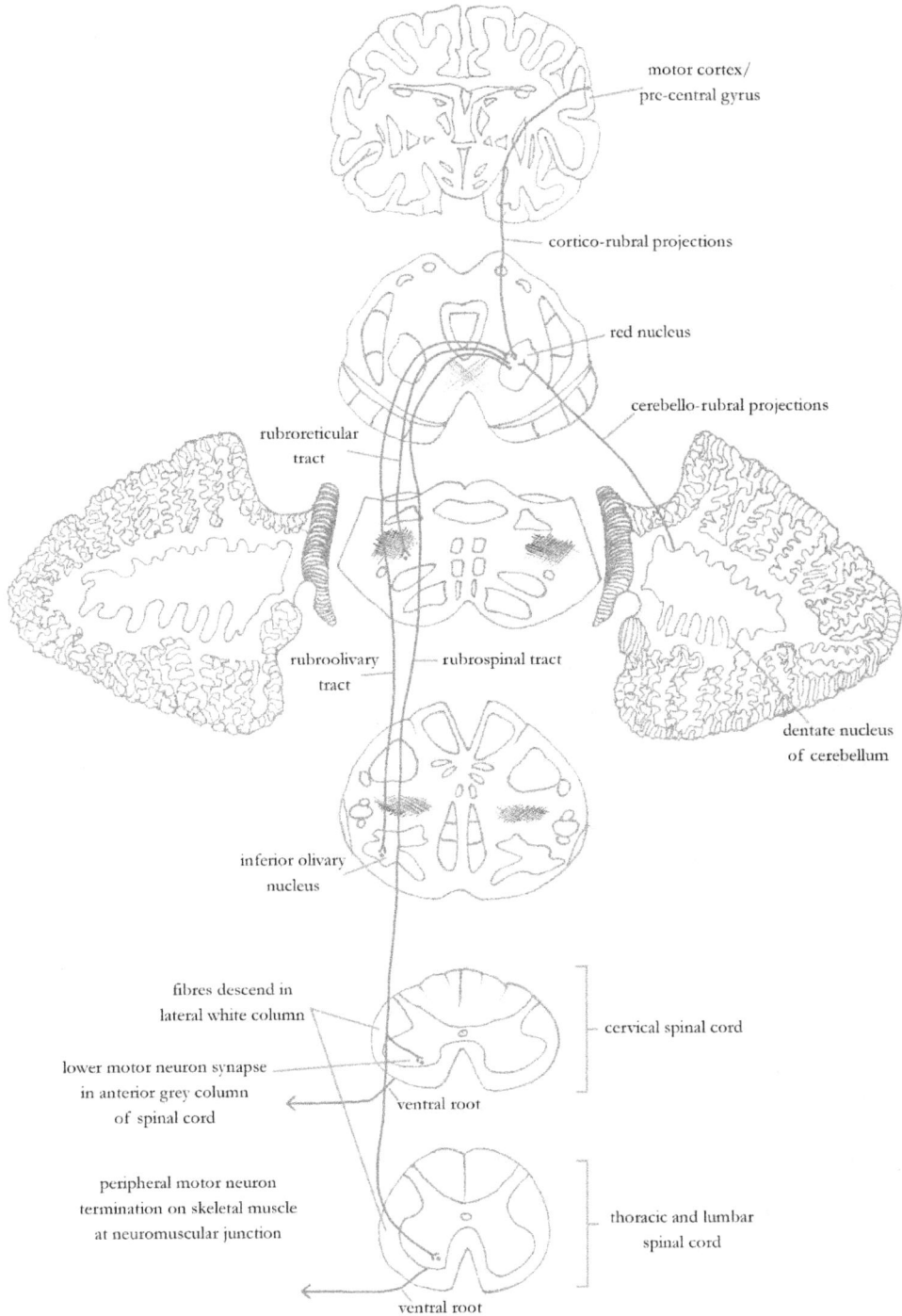

motor cortex/
pre-central gyrus

cortico-rubral projections

red nucleus

cerebello-rubral projections

rubroreticular
tract

rubroolivary
tract

rubrospinal tract

dentate nucleus
of cerebellum

inferior olivary
nucleus

fibres descend in
lateral white column

cervical spinal cord

lower motor neuron synapse
in anterior grey column
of spinal cord

ventral root

peripheral motor neuron
termination on skeletal muscle
at neuromuscular junction

thoracic and lumbar
spinal cord

ventral root

Medial & Lateral Vestibulospinal Tract
With Cerebellar Projections and Input From CNVIII

Facilitates activity of extensor muscles and inhibits flexor muscles

CEREBELLAR PROJECTIONS & INPUT FROM INNER EAR VESTIBULOCOCHLEAR NERVE
(Upper motor neuron)

MEDIAL VESTIBULAR NUECLEI IN PONS AND MEDULLA OBLONGATA	LATERAL VESTIBULAR NURCLEI IN PONS AND MEDULLA OBLONGATA
FIBRES TRAVEL THROUGH MEDIAL LONGITUDINAL FASCICULUS OF MEDULLA OBLONGATA	FIBRES TRAVEL THROUGH VENTEROLATERAL AREAS OF PONS AND MEDULLA OBLONGATA
ANTERIOR WHITE COLUMN OF CERVICAL SPINAL CORD	VENTEROLATERAL WHITE COLUMN OF LUMBAR SPINAL CORD
ANTERIOR GREY COLUMN OF CERVICAL SPINAL CORD (Synapse to lower motor neuron)	ANTERIOR GREY COLUMN OF LUMBAR SPINAL CORD (Synapse to lower motor neuron)
VENTRAL ROOT	VENTRAL ROOT
PERIPHERAL MOTOR NERVE	PERIPHERAL MOTOR NERVE
TERMINATION ON UPPER TRUNK AND UPPER LIMB SKELETAL MUSCLE AT NEUROMUSCULAR JUNCTION	TERMINATION ON LOWER TRUNK AND LOWER LIMB SKELETAL MUSCLE AT NEUROMUSCULAR JUNCTION

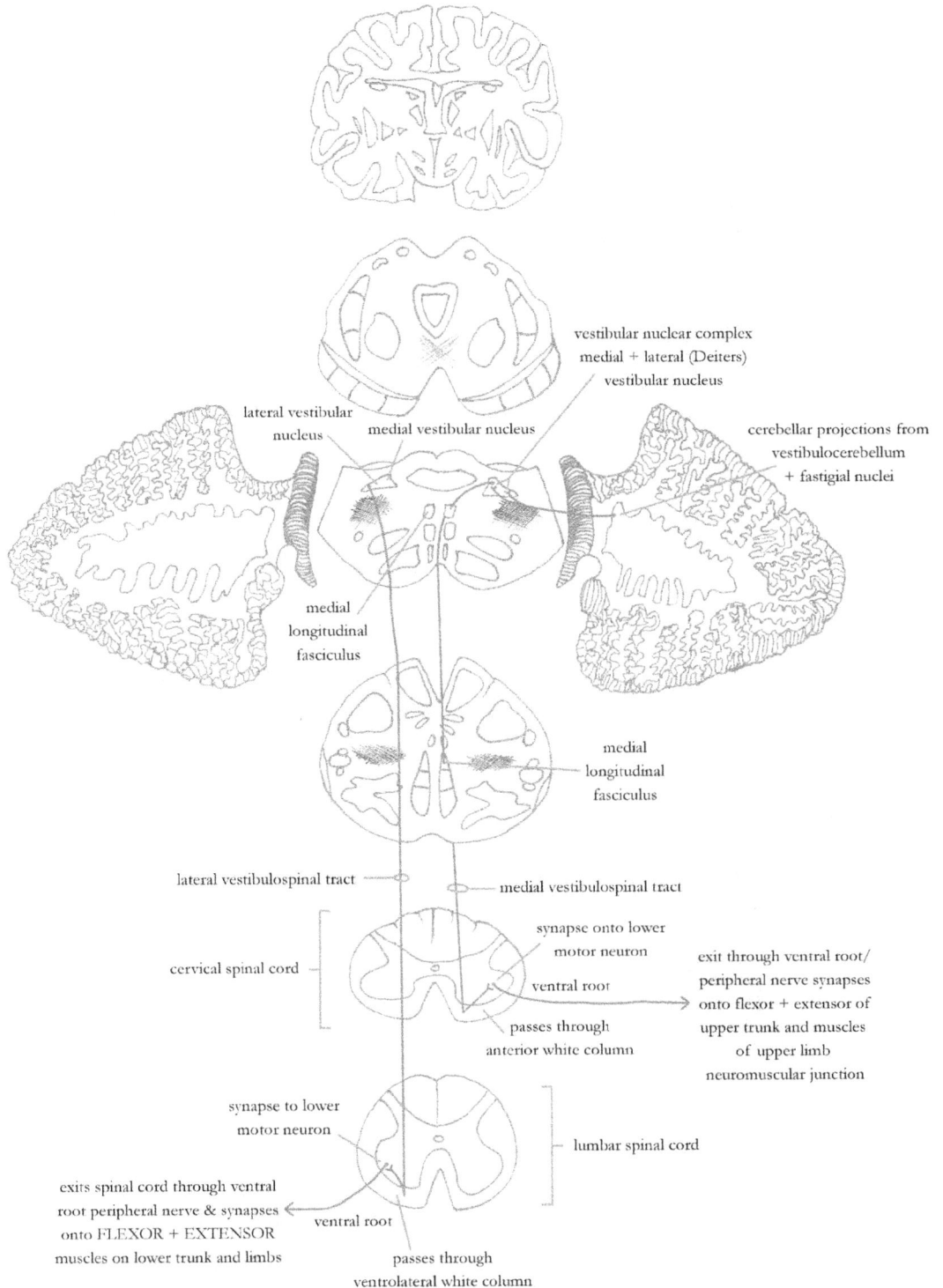

vestibular nuclear complex
medial + lateral (Deiters)
vestibular nucleus

lateral vestibular
nucleus

medial vestibular nucleus

cerebellar projections from
vestibulocerebellum
+ fastigial nuclei

medial
longitudinal
fasciculus

medial
longitudinal
fasciculus

lateral vestibulospinal tract

medial vestibulospinal tract

cervical spinal cord

synapse onto lower
motor neuron

ventral root

passes through
anterior white column

exit through ventral root/
peripheral nerve synapses
onto flexor + extensor of
upper trunk and muscles
of upper limb
neuromuscular junction

synapse to lower
motor neuron

lumbar spinal cord

exits spinal cord through ventral
root peripheral nerve & synapses
onto FLEXOR + EXTENSOR
muscles on lower trunk and limbs

ventral root

passes through
ventrolateral white column

Olivospinal Tract

Unconscious proprioception

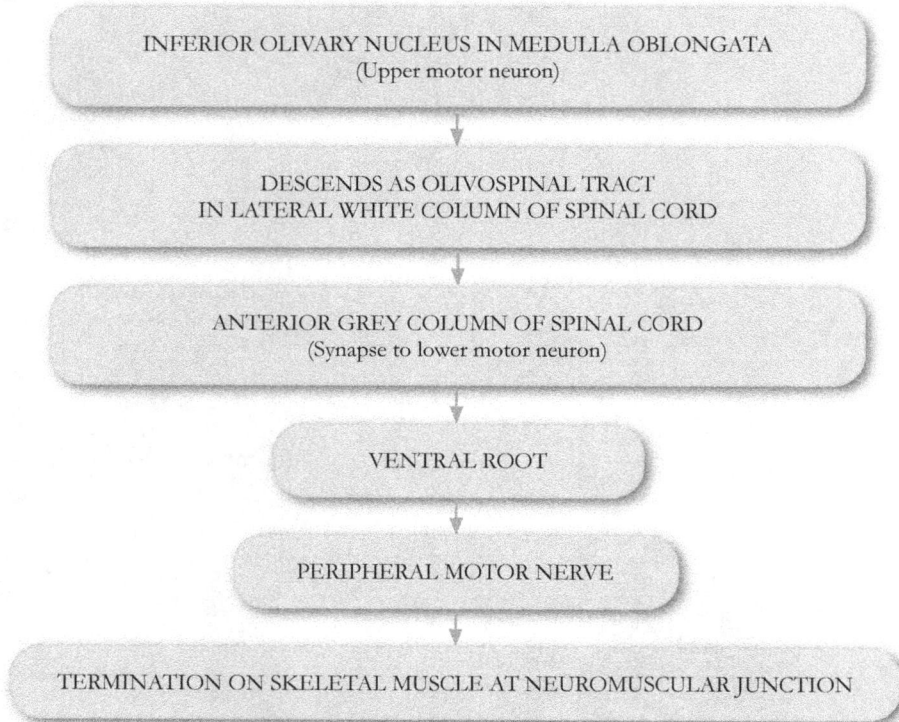

INFERIOR OLIVARY NUCLEUS IN MEDULLA OBLONGATA
(Upper motor neuron)

DESCENDS AS OLIVOSPINAL TRACT
IN LATERAL WHITE COLUMN OF SPINAL CORD

ANTERIOR GREY COLUMN OF SPINAL CORD
(Synapse to lower motor neuron)

VENTRAL ROOT

PERIPHERAL MOTOR NERVE

TERMINATION ON SKELETAL MUSCLE AT NEUROMUSCULAR JUNCTION

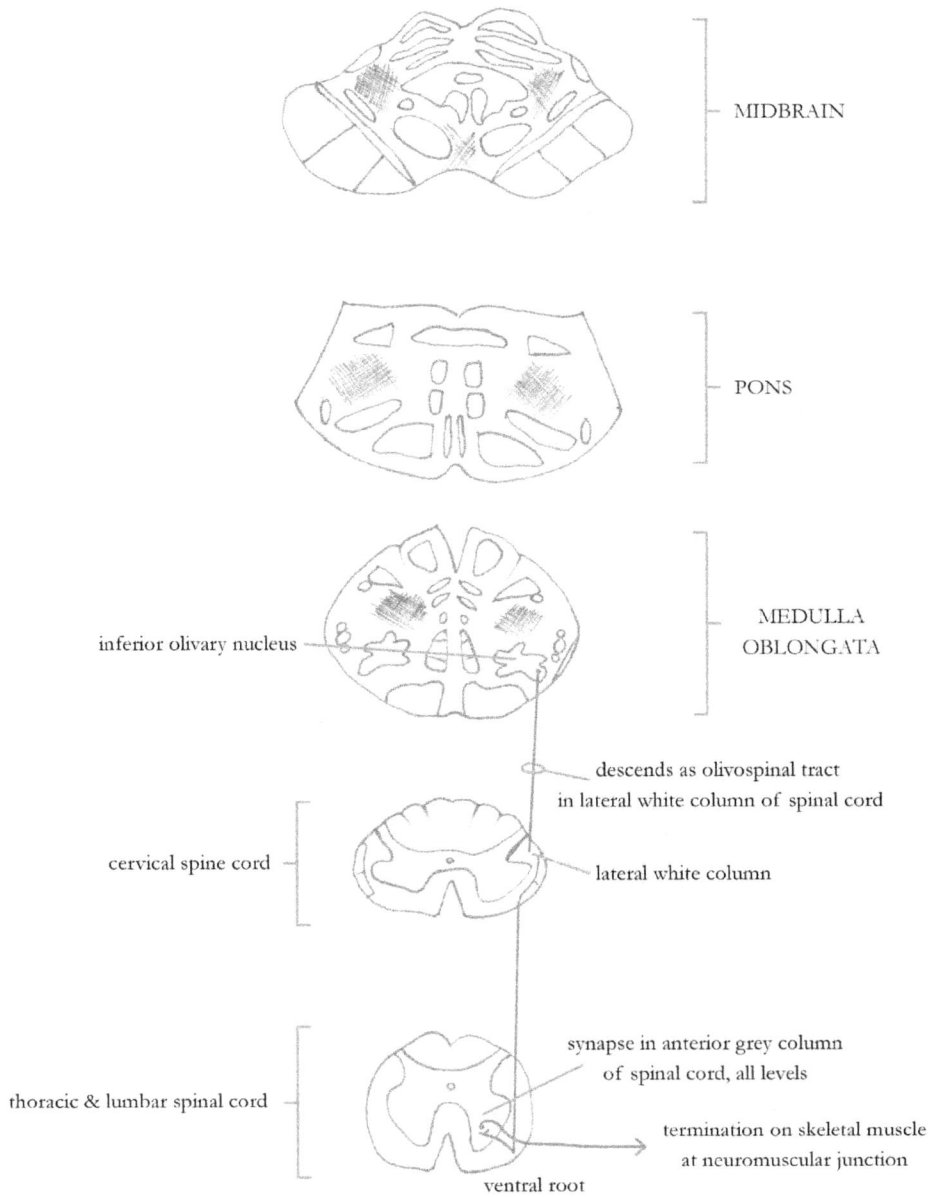

MIDBRAIN

PONS

MEDULLA
OBLONGATA

inferior olivary nucleus

descends as olivospinal tract
in lateral white column of spinal cord

cervical spine cord

lateral white column

synapse in anterior grey column
of spinal cord, all levels

thoracic & lumbar spinal cord

termination on skeletal muscle
at neuromuscular junction

ventral root

Interstitiospinal Tract

Movement of the head and eyes for the control of vertical and rotatory gaze shifts

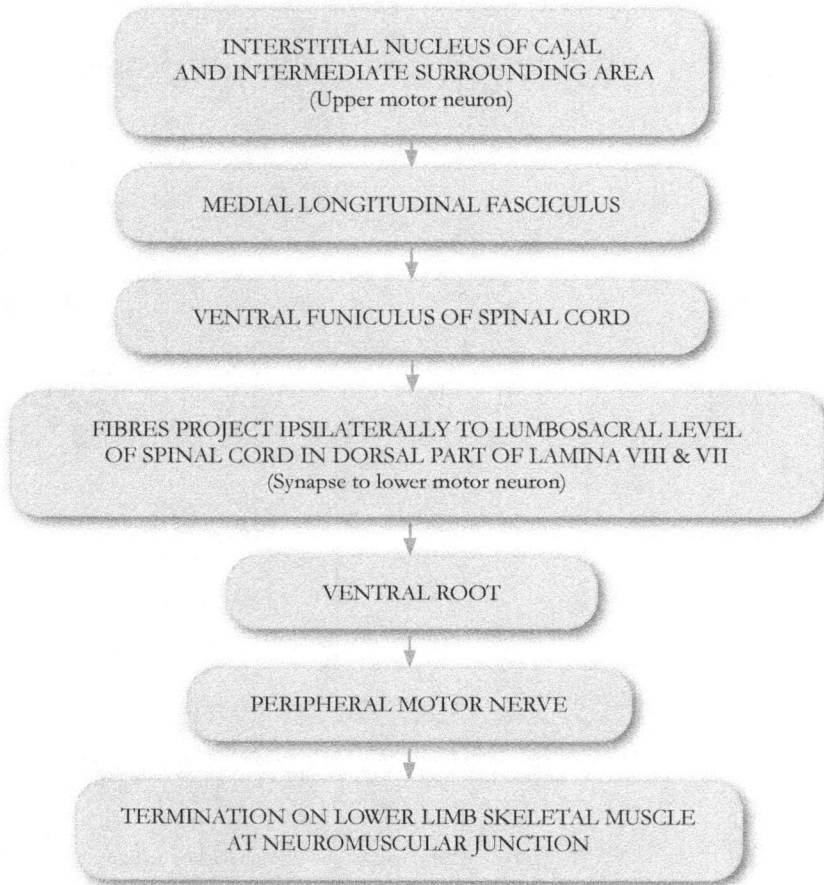

INTERSTITIAL NUCLEUS OF CAJAL
AND INTERMEDIATE SURROUNDING AREA
(Upper motor neuron)

↓

MEDIAL LONGITUDINAL FASCICULUS

↓

VENTRAL FUNICULUS OF SPINAL CORD

↓

FIBRES PROJECT IPSILATERALLY TO LUMBOSACRAL LEVEL
OF SPINAL CORD IN DORSAL PART OF LAMINA VIII & VII
(Synapse to lower motor neuron)

↓

VENTRAL ROOT

↓

PERIPHERAL MOTOR NERVE

↓

TERMINATION ON LOWER LIMB SKELETAL MUSCLE
AT NEUROMUSCULAR JUNCTION

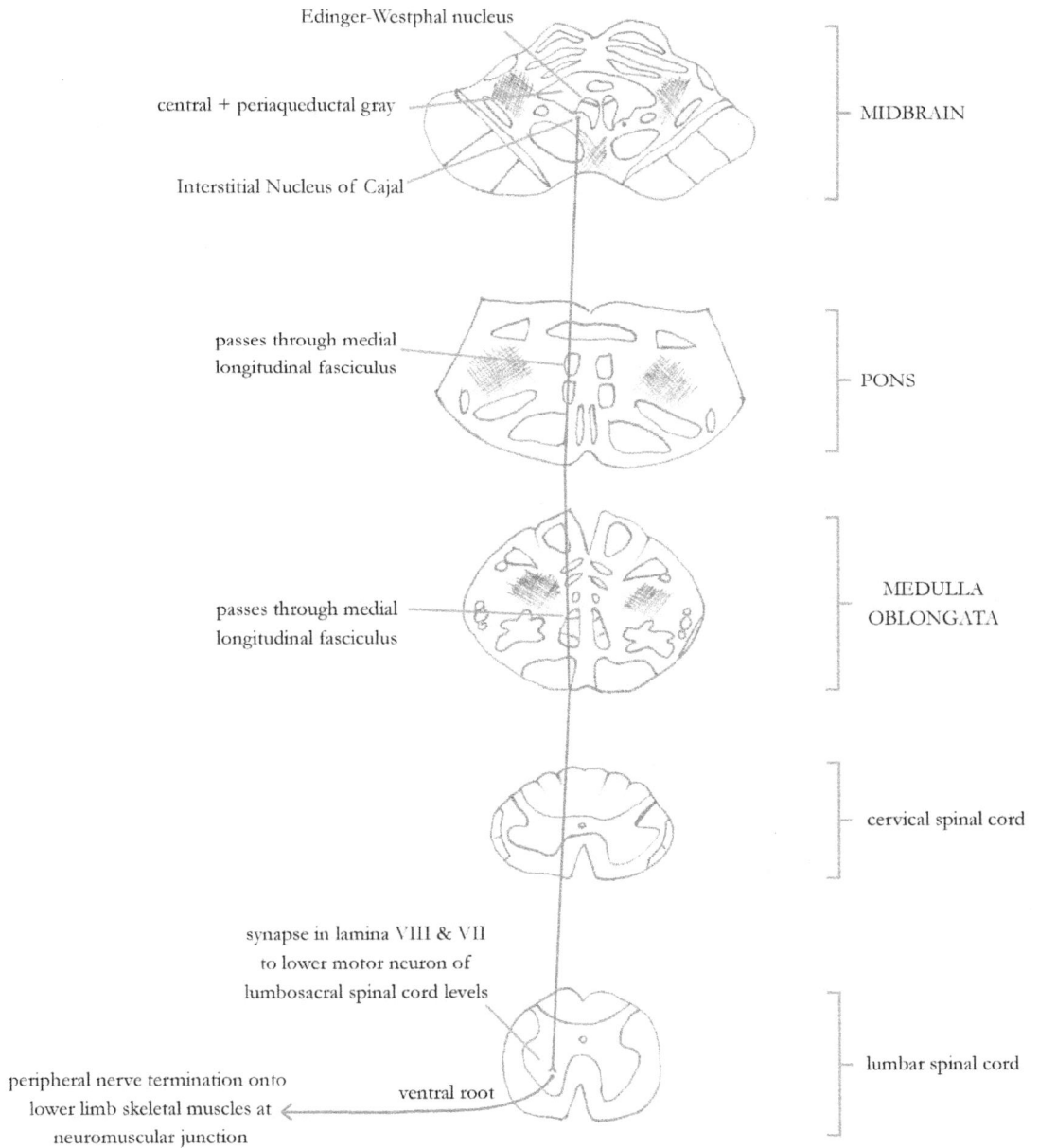

Edinger-Westphal nucleus

central + periaqueductal gray

Interstitial Nucleus of Cajal

MIDBRAIN

passes through medial
longitudinal fasciculus

PONS

passes through medial
longitudinal fasciculus

MEDULLA
OBLONGATA

cervical spinal cord

synapse in lamina VIII & VII
to lower motor neuron of
lumbosacral spinal cord levels

lumbar spinal cord

peripheral nerve termination onto
lower limb skeletal muscles at
neuromuscular junction

ventral root

Solitariospinial Tract
With Visceral Afferent Input From CNVII, IX & X
Integrative control of breathing

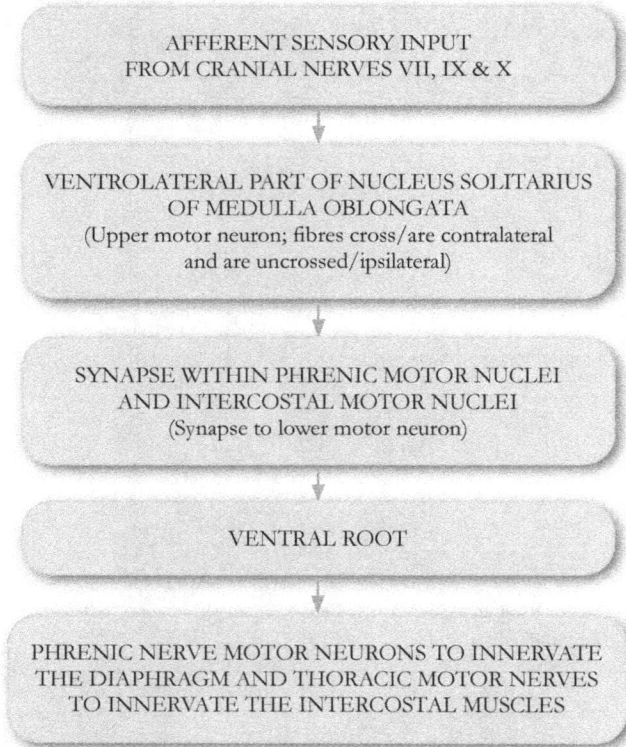

AFFERENT SENSORY INPUT
FROM CRANIAL NERVES VII, IX & X

↓

VENTROLATERAL PART OF NUCLEUS SOLITARIUS
OF MEDULLA OBLONGATA
(Upper motor neuron; fibres cross/are contralateral
and are uncrossed/ipsilateral)

↓

SYNAPSE WITHIN PHRENIC MOTOR NUCLEI
AND INTERCOSTAL MOTOR NUCLEI
(Synapse to lower motor neuron)

↓

VENTRAL ROOT

↓

PHRENIC NERVE MOTOR NEURONS TO INNERVATE
THE DIAPHRAGM AND THORACIC MOTOR NERVES
TO INNERVATE THE INTERCOSTAL MUSCLES

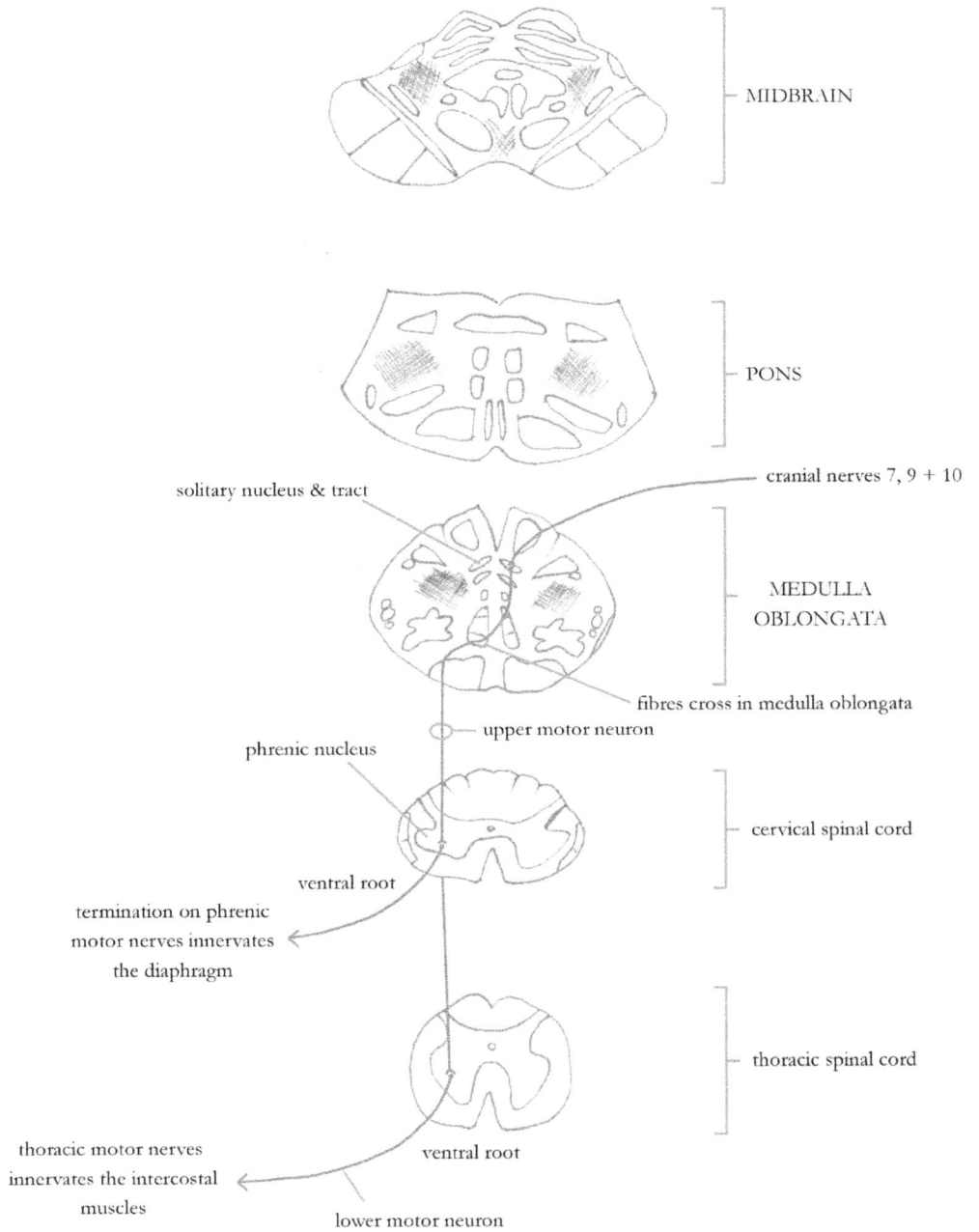

MIDBRAIN

PONS

cranial nerves 7, 9 + 10

solitary nucleus & tract

MEDULLA
OBLONGATA

fibres cross in medulla oblongata

upper motor neuron

phrenic nucleus

cervical spinal cord

ventral root

termination on phrenic
motor nerves innervates
the diaphragm

thoracic spinal cord

thoracic motor nerves
innervates the intercostal
muscles

ventral root

lower motor neuron

Descending Autonomic Pathway Sympathetic

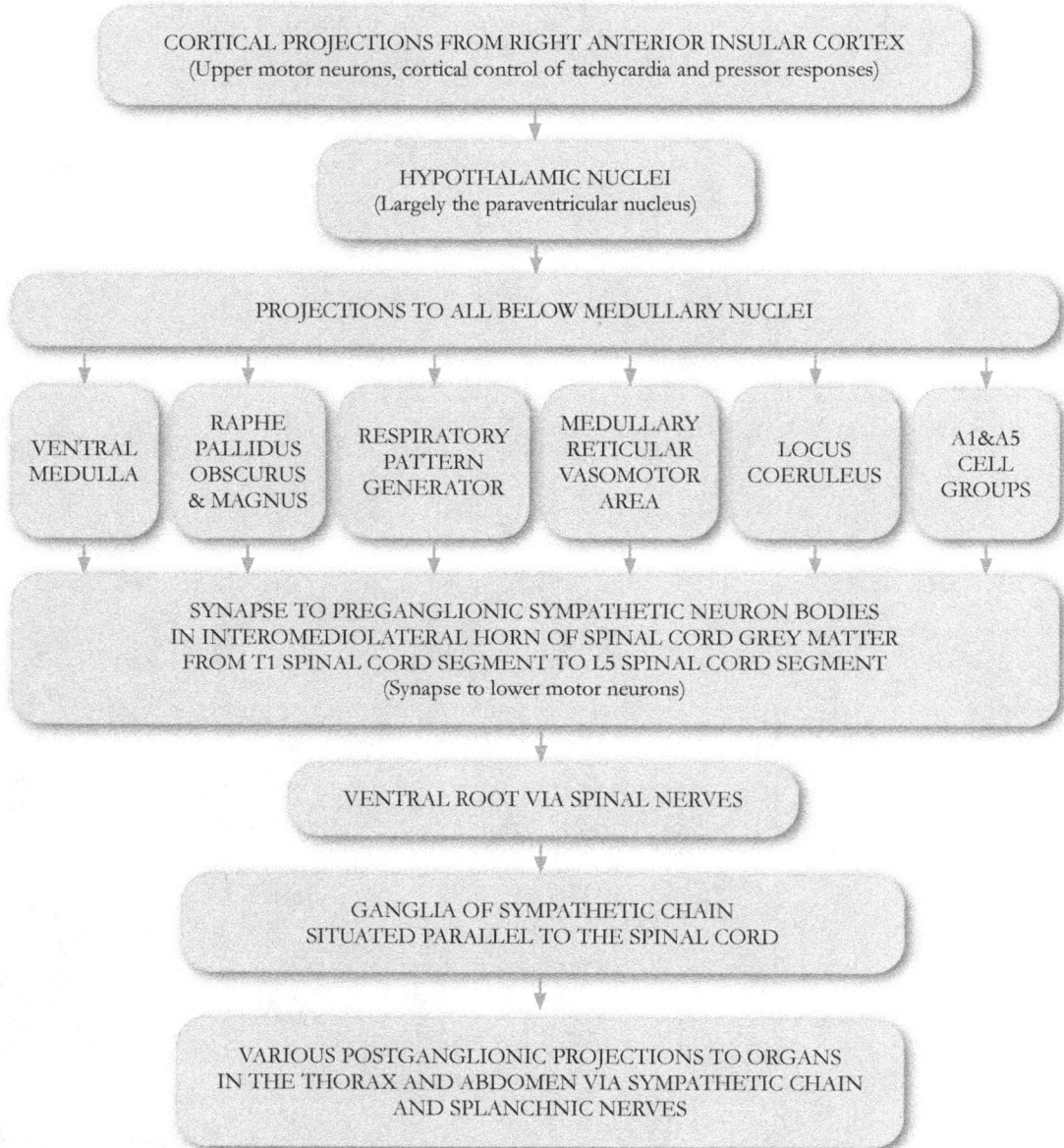

CORTICAL PROJECTIONS FROM RIGHT ANTERIOR INSULAR CORTEX
(Upper motor neurons, cortical control of tachycardia and pressor responses)

HYPOTHALAMIC NUCLEI
(Largely the paraventricular nucleus)

PROJECTIONS TO ALL BELOW MEDULLARY NUCLEI

| VENTRAL MEDULLA | RAPHE PALLIDUS OBSCURUS & MAGNUS | RESPIRATORY PATTERN GENERATOR | MEDULLARY RETICULAR VASOMOTOR AREA | LOCUS COERULEUS | A1&A5 CELL GROUPS |

SYNAPSE TO PREGANGLIONIC SYMPATHETIC NEURON BODIES
IN INTEROMEDIOLATERAL HORN OF SPINAL CORD GREY MATTER
FROM T1 SPINAL CORD SEGMENT TO L5 SPINAL CORD SEGMENT
(Synapse to lower motor neurons)

VENTRAL ROOT VIA SPINAL NERVES

GANGLIA OF SYMPATHETIC CHAIN
SITUATED PARALLEL TO THE SPINAL CORD

VARIOUS POSTGANGLIONIC PROJECTIONS TO ORGANS
IN THE THORAX AND ABDOMEN VIA SYMPATHETIC CHAIN
AND SPLANCHNIC NERVES

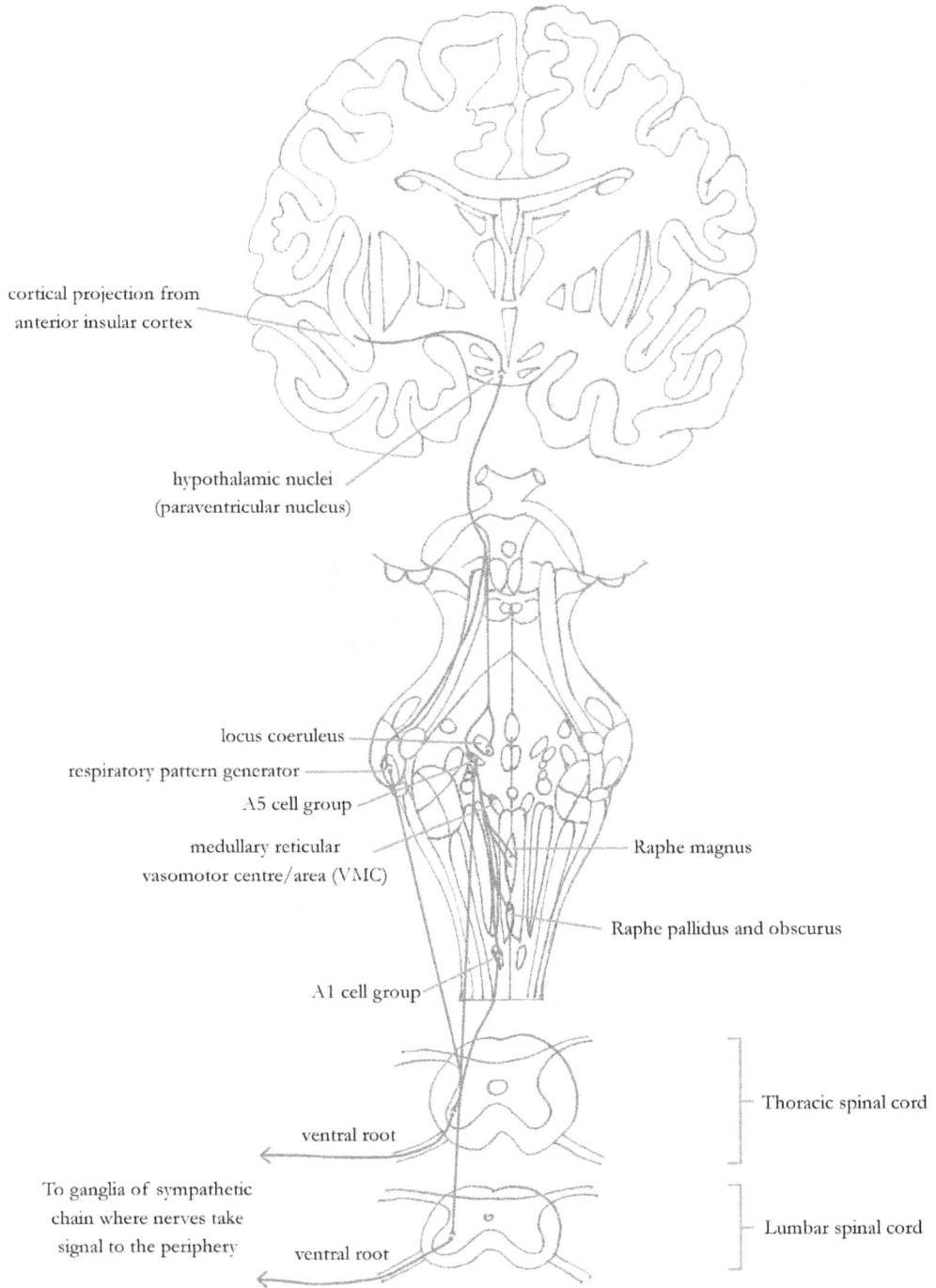

cortical projection from
anterior insular cortex

hypothalamic nuclei
(paraventricular nucleus)

locus coeruleus

respiratory pattern generator

A5 cell group

medullary reticular
vasomotor centre/area (VMC)

Raphe magnus

Raphe pallidus and obscurus

A1 cell group

Thoracic spinal cord

ventral root

Lumbar spinal cord

To ganglia of sympathetic
chain where nerves take
signal to the periphery

ventral root

MOTOR, SENSORY & PARASYMPATHETIC PATHWAYS OF CRANIAL NERVES

The following flow diagrams are the central pathways of corticobulbar/corticonuclear tracts. General considerations:

1. Each cranial nerve is either sensory, motor, or mixed and may have a parasympathetic component.

2. Somatic or branchial upper motor neurons project from the cortex to the brainstem nuclei, where they synapse onto lower motor neurons.

3. Somatic or branchial lower motor neurons are the cranial nerves that project from brainstem nuclei to the periphery.

4. Sensory cranial nerve pathways are composed of a 1st order, 2nd order and 3rd order neuron. 1st order neurons project from the periphery to the brainstem nuclei, 2nd order neurons project from the brainstem nuclei to a nucleus within the thalamus, and 3rd order neurons project from a nucleus in the thalamus to an area of the cerebral cortex.

5. Parasympathetic efferent cranial nerve pathways are composed of an upper motor neuron, a lower motor neuron and a tertiary neuron. Upper motor neurons project from the cortex to the brainstem nuclei, lower motor neurons project from the brainstem nuclei to a peripheral ganglion, and tertiary neurons project from the peripheral ganglion to the nerve endings in the peripheral tissues.

Olfactory Nerve CNI

Sensory pathway for smell;
Special visceral afferent

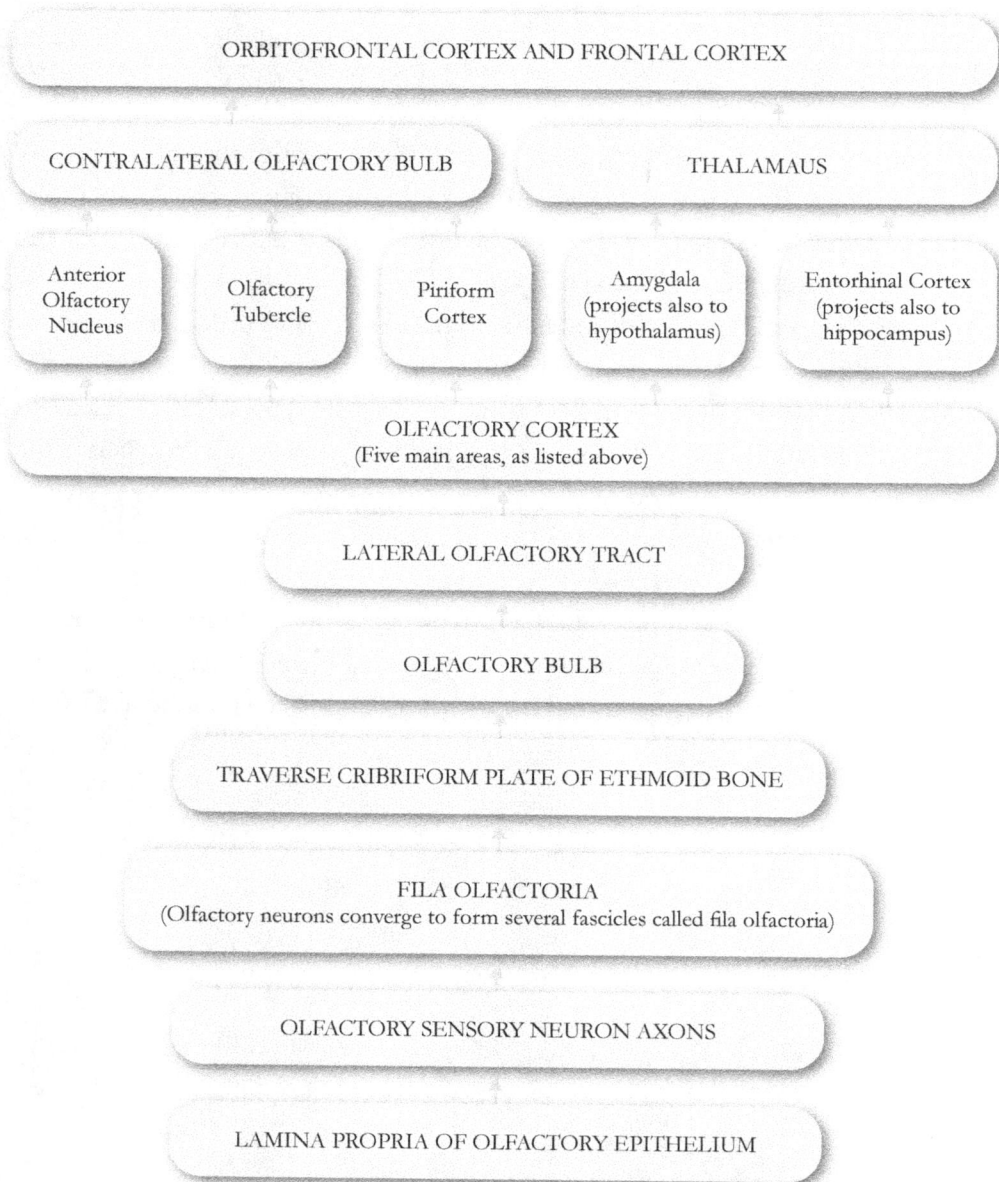

ORBITOFRONTAL CORTEX AND FRONTAL CORTEX

CONTRALATERAL OLFACTORY BULB

THALAMAUS

Anterior Olfactory Nucleus

Olfactory Tubercle

Piriform Cortex

Amygdala (projects also to hypothalamus)

Entorhinal Cortex (projects also to hippocampus)

OLFACTORY CORTEX
(Five main areas, as listed above)

LATERAL OLFACTORY TRACT

OLFACTORY BULB

TRAVERSE CRIBRIFORM PLATE OF ETHMOID BONE

FILA OLFACTORIA
(Olfactory neurons converge to form several fascicles called fila olfactoria)

OLFACTORY SENSORY NEURON AXONS

LAMINA PROPRIA OF OLFACTORY EPITHELIUM

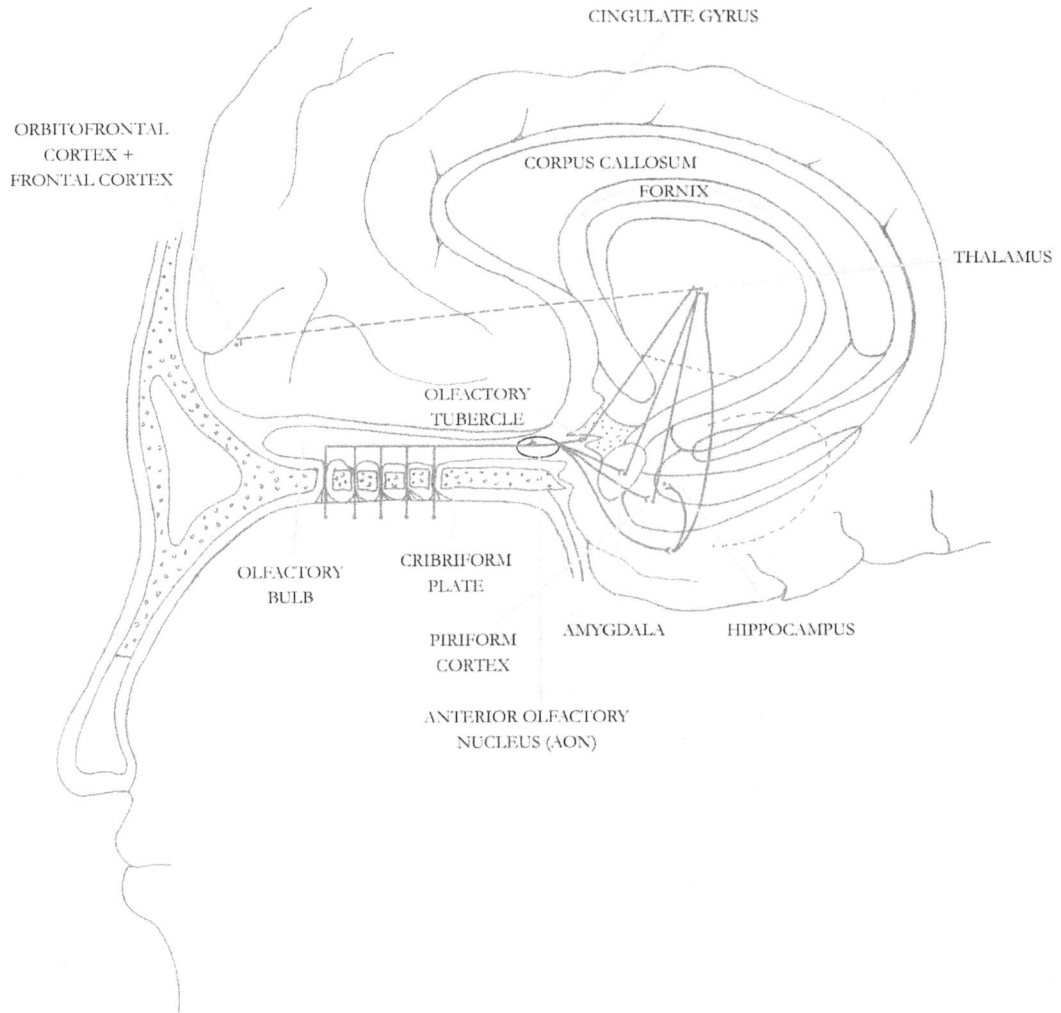

CINGULATE GYRUS

ORBITOFRONTAL
CORTEX +
FRONTAL CORTEX

CORPUS CALLOSUM

FORNIX

THALAMUS

OLFACTORY
TUBERCLE

OLFACTORY
BULB

CRIBRIFORM
PLATE

PIRIFORM
CORTEX

AMYGDALA

HIPPOCAMPUS

ANTERIOR OLFACTORY
NUCLEUS (AON)

Optic Nerve CNII

Sensory pathway for vision;
Special somatic afferent

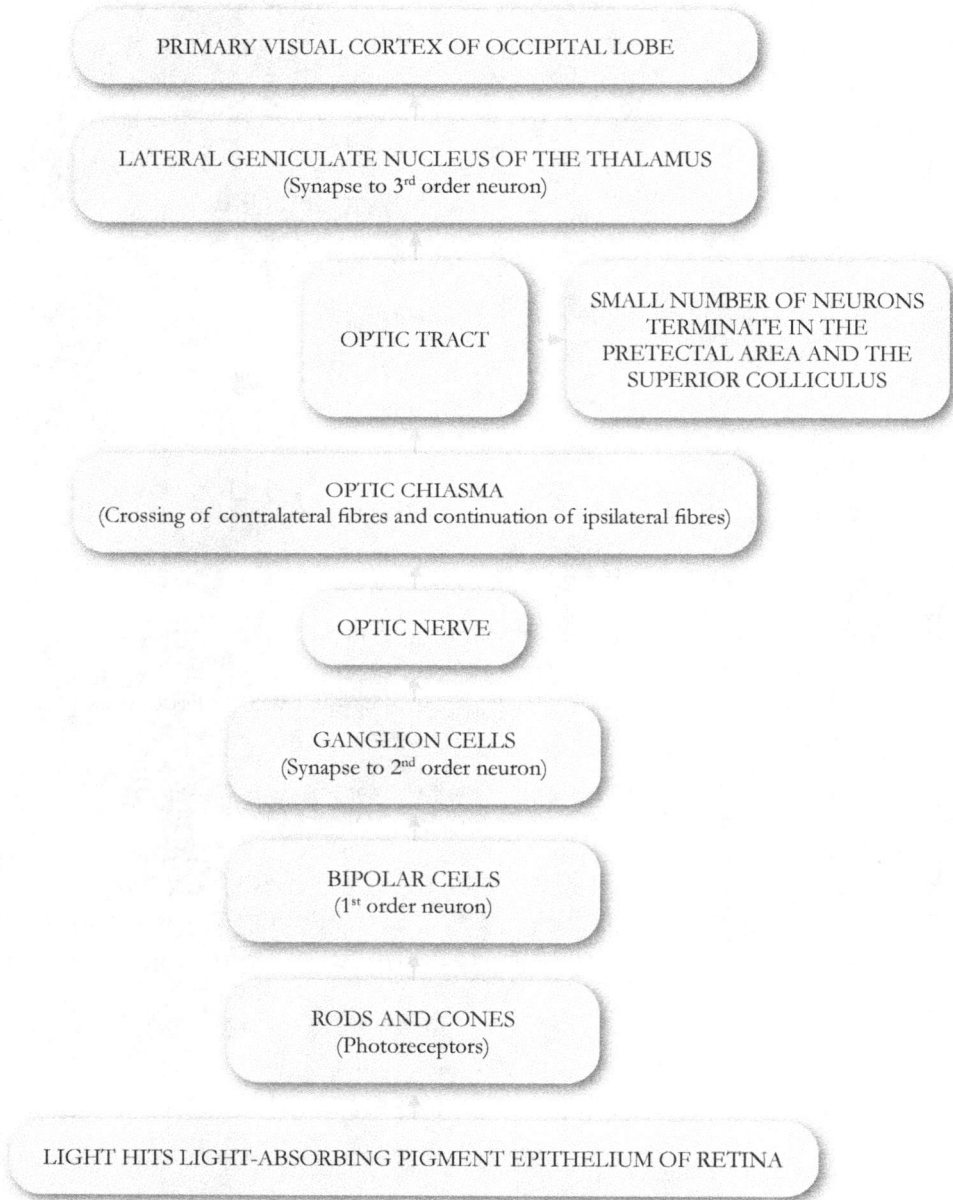

PRIMARY VISUAL CORTEX OF OCCIPITAL LOBE

LATERAL GENICULATE NUCLEUS OF THE THALAMUS
(Synapse to 3rd order neuron)

OPTIC TRACT

SMALL NUMBER OF NEURONS TERMINATE IN THE PRETECTAL AREA AND THE SUPERIOR COLLICULUS

OPTIC CHIASMA
(Crossing of contralateral fibres and continuation of ipsilateral fibres)

OPTIC NERVE

GANGLION CELLS
(Synapse to 2nd order neuron)

BIPOLAR CELLS
(1st order neuron)

RODS AND CONES
(Photoreceptors)

LIGHT HITS LIGHT-ABSORBING PIGMENT EPITHELIUM OF RETINA

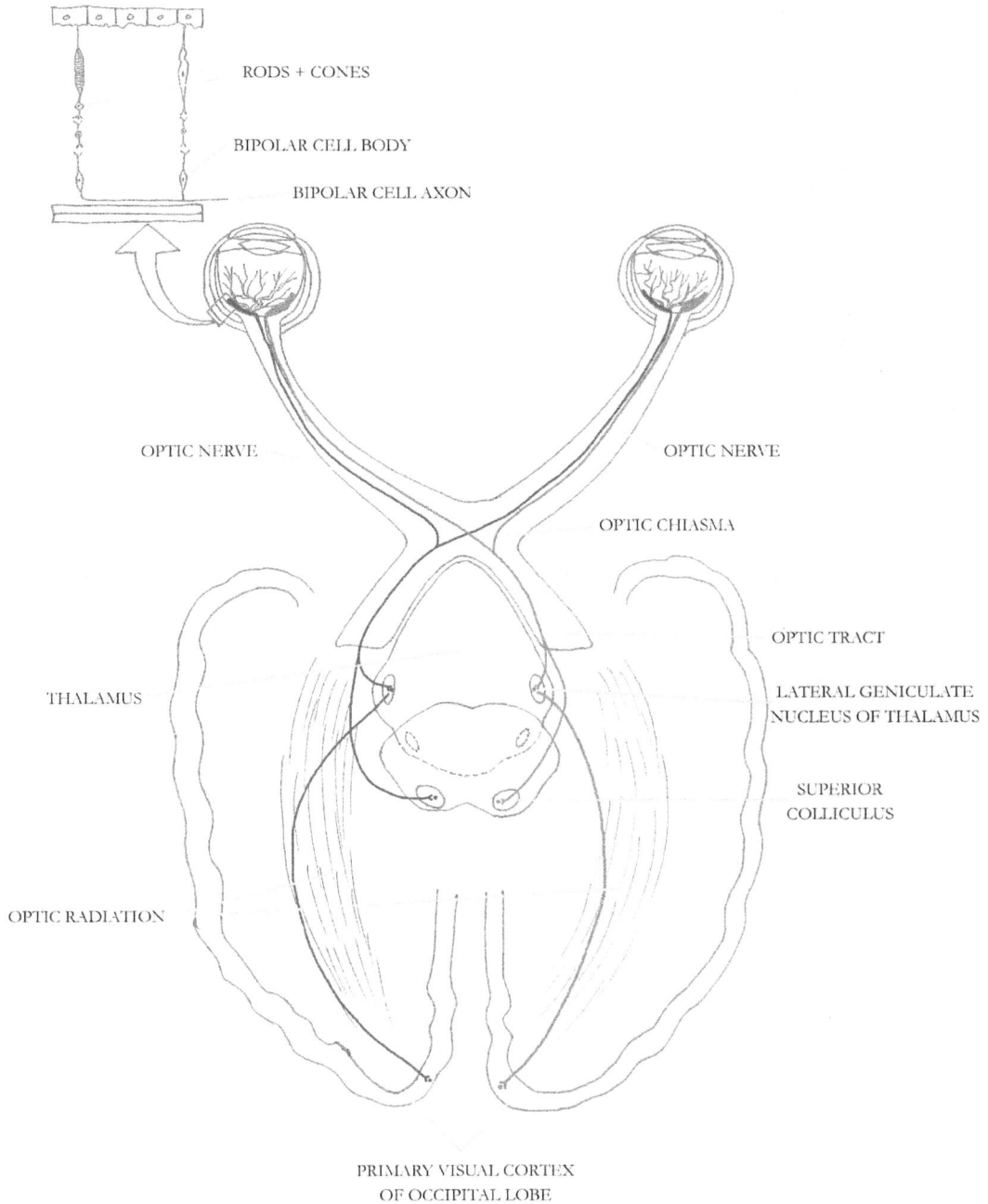

RODS + CONES

BIPOLAR CELL BODY

BIPOLAR CELL AXON

OPTIC NERVE

OPTIC NERVE

OPTIC CHIASMA

OPTIC TRACT

THALAMUS

LATERAL GENICULATE
NUCLEUS OF THALAMUS

SUPERIOR
COLLICULUS

OPTIC RADIATION

PRIMARY VISUAL CORTEX
OF OCCIPITAL LOBE

Oculomotor Nerve CNIII

General somatic efferent (motor) fibres to extrinsic eye muscles;
General visceral efferent (parasympathetic) fibres to ciliary ganglion

FIBRES PROJECT FROM THE CORTEX, DENTATE NUCLEUS OF CEREBELLUM, INTERSTITIAL NUCLEUS OF CAJAL, SUPERIOR COLLICULUS, MEDIAL LONGITUDINAL FASCICULUS, CAUDATE NUCLEUS & SUBSTANTIA NIGRA, TROCHLEAR, ABDUCENS & VESTIBULAR NUCLEI

FIBRES PROJECT FROM THE PRETECTAL NUCLEI (MAINLY PRETECTAL OLIVARY NUCLEI), THE VISUAL CORTEX (CONTROLLING ACCOMMODATION)

OCULOMOTOR NUCLEI OF MIDBRAIN
(Paired nuclei located at the base of periaqueductal grey matter of midbrain, at the level of the superior colliculus)

(Lateral oculomotor subnucleus supplies the ipsilateral inferior and medial rectus, and inferior oblique; the medial subnucleus supplies the contralateral superior rectus; the central subnucleus supplies levator palpebrae superioris bilaterally)

EDINGER-WESTPHAL NUCLEUS
(Paired nuclei, located posterior to the oculomotor nuclei, at the level of the superior colliculus/parasympathetic nucleus which supplies the ciliary ganglion for pupillary light reflex)

OCULOMOTOR NERVE

SUPERIOR BRANCH OF OCULOMOTOR NERVE
(Supplies superior rectus muscle, and levator palpebrae superioris)

INFERIOR BRANCH OF OCULOMOTOR NERVE
(Supplies medial and inferior rectus, and inferior oblique muscles, parasympathetic innervation to ciliary ganglion which supplies sphincter pupillae and the ciliary muscle)

caudate nucleus

Frontal eye field Brodmann area 8
(saccadic movement + voluntary
eye movement)
+ motor cortex on precentral
gyrus (voluntary eye movement)
+ supplimentary eye fields

substantia nigra

Edinger-Westphal nucleus

pretectal olivary nucleus

oculomotor nucleus

oculomotor nerve carrying
fibres of general somatic
efferent and general
visceral efferent

superior colliculus

trochlear nucleus

abducens nucleus

medial vestibular
nucleus

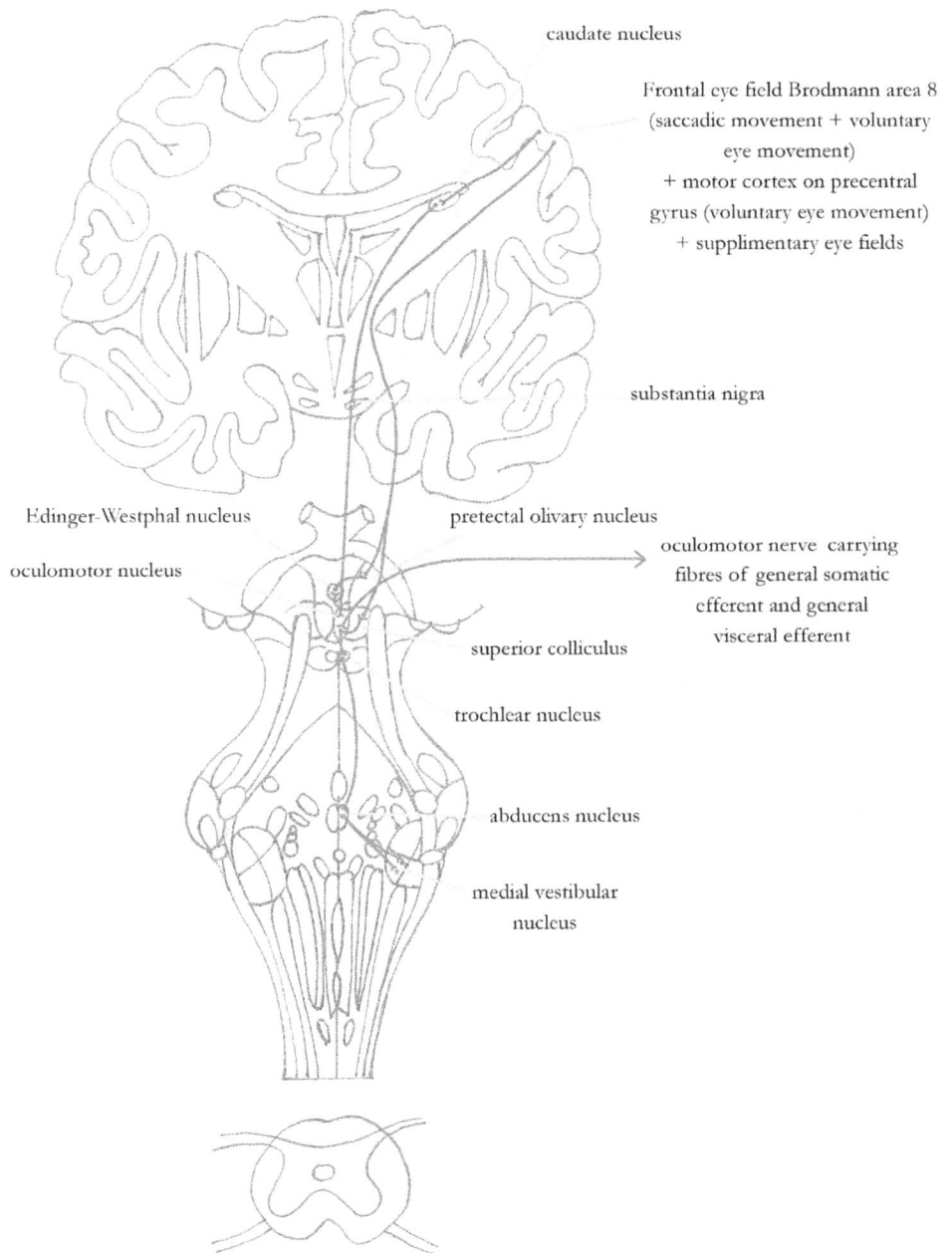

* This diagram loosly illustrates the connection during various eye movements, such as saccadic eye movement, smooth pursuit, gaze, active visual fixation and vergence. It is beyond the scope of this book to discuss the neurophysiology of such control systems.

Trochlear Nerve CNIV

Somatic motor to superior oblique muscle of eye;
General somatic efferent

FIBRES PROJECT FROM THE CORTEX (FRONTAL EYE FIELDS, SUPPLEMENTARY
EYE FIELDS, & POSTERIOR PARIETAL CORTEX), DENTATE NUCLEUS OF
CEREBELLUM, INTERSTITIAL NUCLEUS OF CAJAL, CAUDATE NUCLEUS, SUBSTANTIA NIGRA,
SUPERIOR COLLICULUS, MEDIAL LONGITUDINAL FASCICULUS

TROCHLEAR NUCLEUS
(Lies in periaqueductal gray, fibres cross the midline;
synapse here from upper motor neurons to lower motor neurons)

TROCHLEAR NERVE

NEUROMUSCULAR JUNCTION OF
SUPERIOR OBLIQUE MUSCLE

CORTEX OF FRONTAL EYE FIELDS
+ supplimentary eye fields
+ posterior parietal cortex
connected with caudate nucleus

DENTATE NUCLEUS
OF CEREBELLUM

Interstitial nucleus
of Cajal

from superior colliculus

TROCHLEAR
NERVE

trochlear nucleus

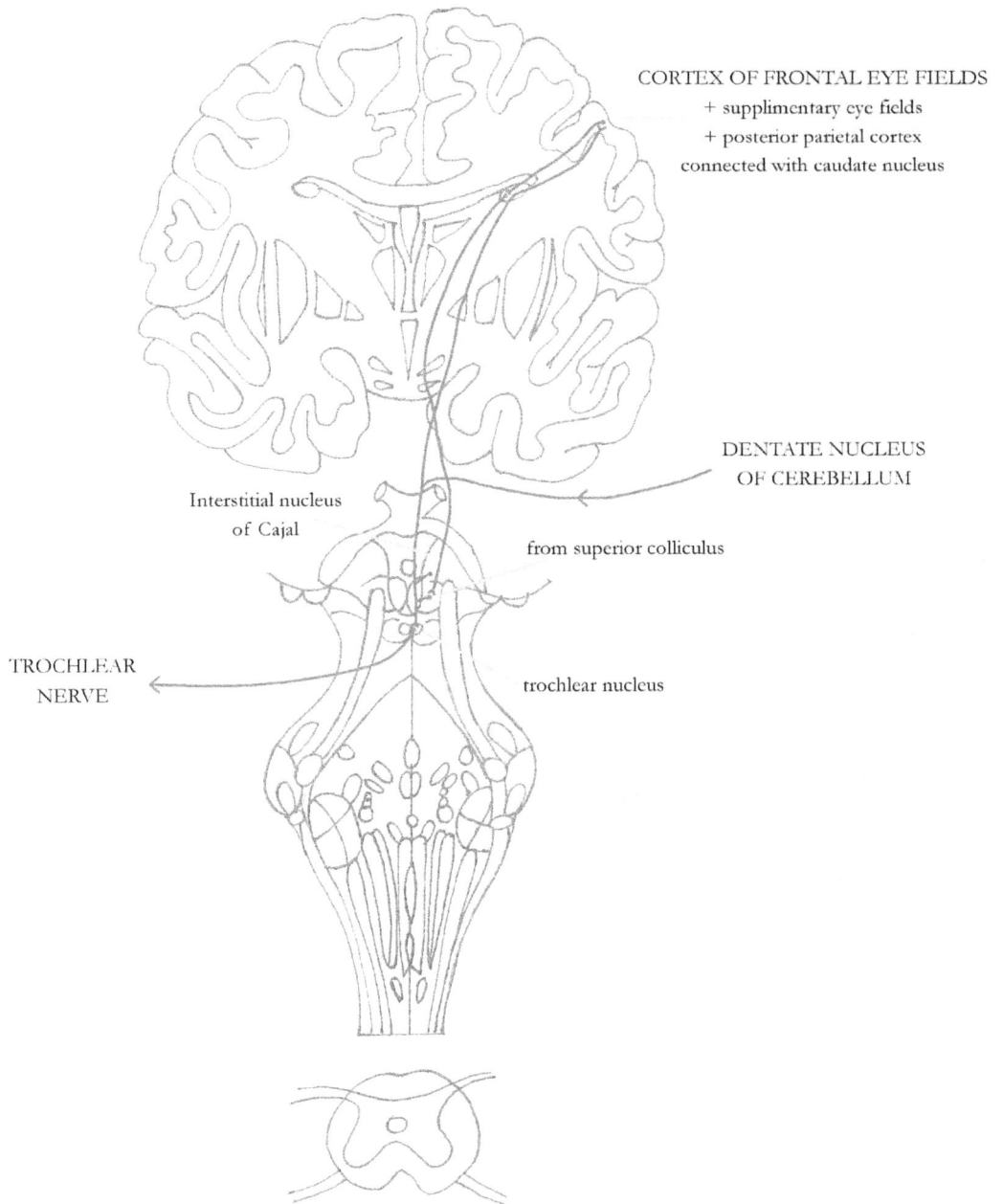

* This image shows some of the connections during saccadic eye movement, smooth pursuit, gaze, active visual fixation and vergence.

Trigeminal Nerve CNV – Branchial Motor Pathway

Movement of muscles of mastication, mylohyoid, anterior belly of digastric, tensor tympani and tensor veli palatini muscles, special visceral efferent

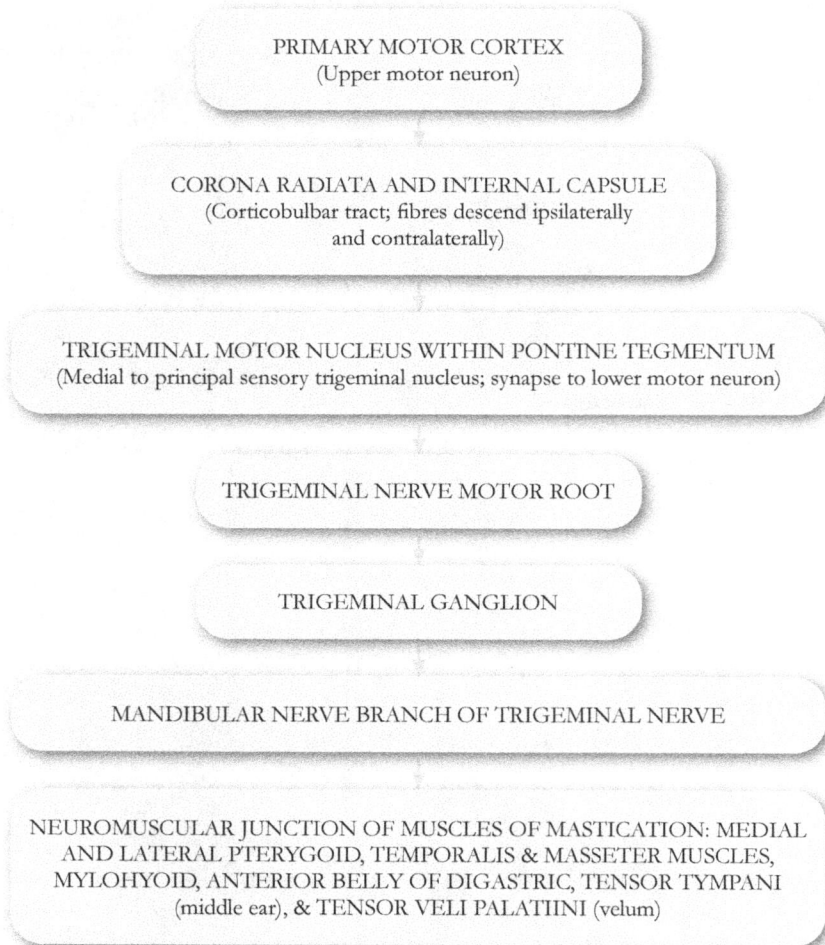

PRIMARY MOTOR CORTEX
(Upper motor neuron)

CORONA RADIATA AND INTERNAL CAPSULE
(Corticobulbar tract; fibres descend ipsilaterally
and contralaterally)

TRIGEMINAL MOTOR NUCLEUS WITHIN PONTINE TEGMENTUM
(Medial to principal sensory trigeminal nucleus; synapse to lower motor neuron)

TRIGEMINAL NERVE MOTOR ROOT

TRIGEMINAL GANGLION

MANDIBULAR NERVE BRANCH OF TRIGEMINAL NERVE

NEUROMUSCULAR JUNCTION OF MUSCLES OF MASTICATION: MEDIAL
AND LATERAL PTERYGOID, TEMPORALIS & MASSETER MUSCLES,
MYLOHYOID, ANTERIOR BELLY OF DIGASTRIC, TENSOR TYMPANI
(middle ear), & TENSOR VELI PALATIINI (velum)

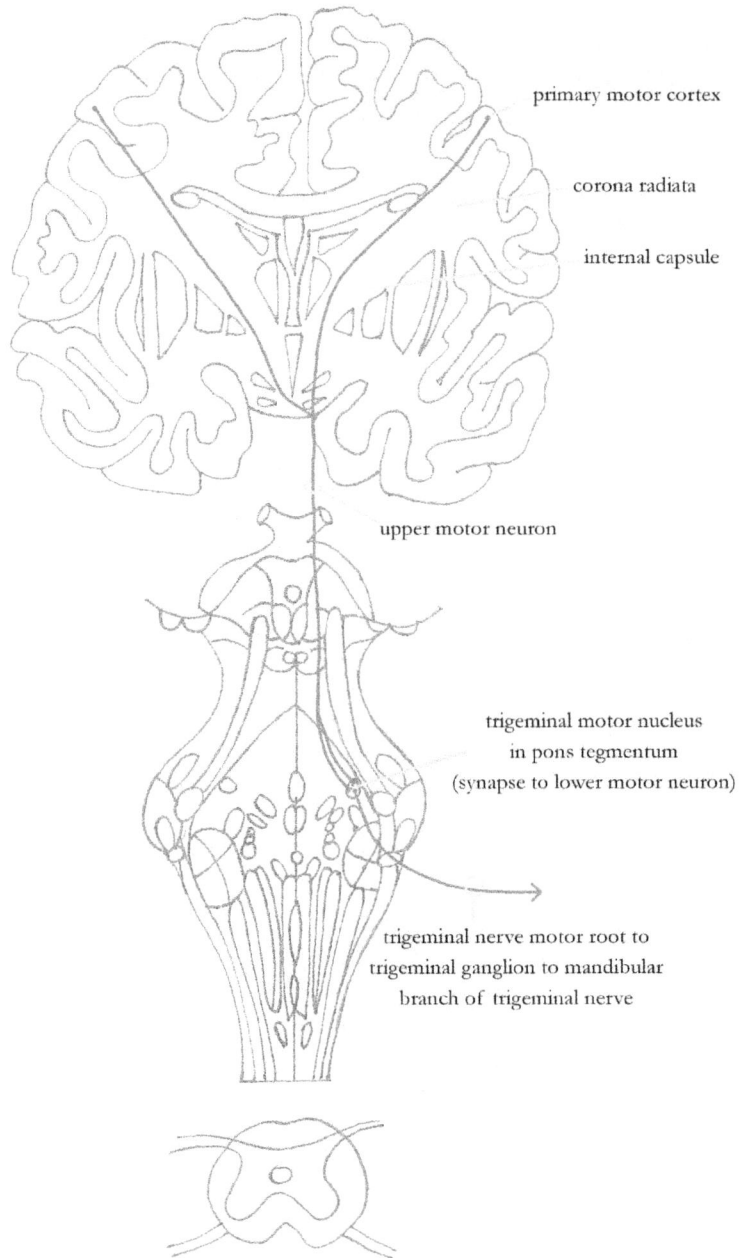

primary motor cortex

corona radiata

internal capsule

upper motor neuron

trigeminal motor nucleus
in pons tegmentum
(synapse to lower motor neuron)

trigeminal nerve motor root to
trigeminal ganglion to mandibular
branch of trigeminal nerve

Trigeminal Nerve CNV – Sensory Pathway

Sensory innervation of face, scalp, cornea, nasal and oral cavities (general sensation of anterior $2/3$ of tongue), cranial dura mater, tentorium cerebelli, posterior area of falx cerebri general somatic afferent

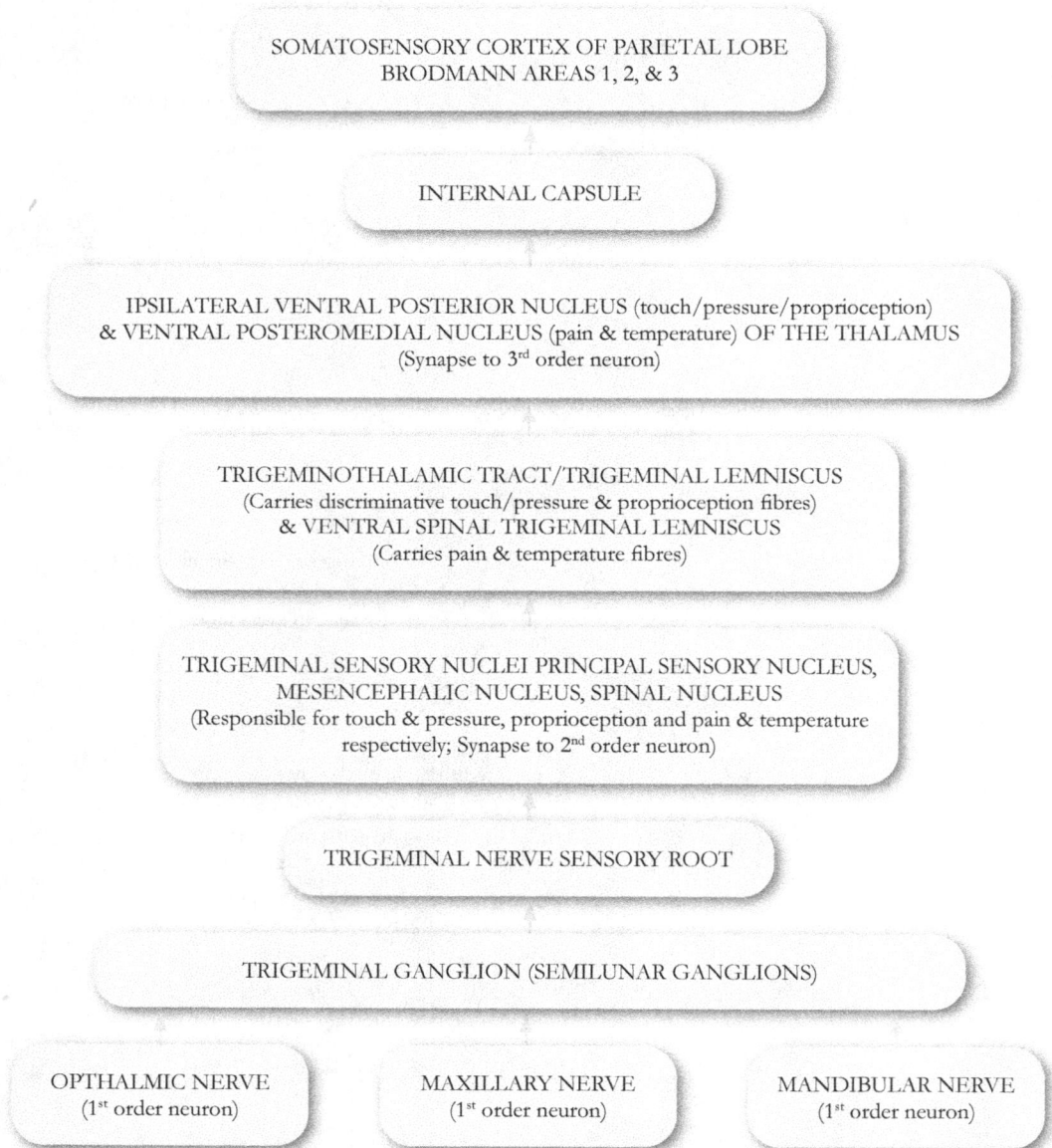

SOMATOSENSORY CORTEX OF PARIETAL LOBE
BRODMANN AREAS 1, 2, & 3

INTERNAL CAPSULE

IPSILATERAL VENTRAL POSTERIOR NUCLEUS (touch/pressure/proprioception)
& VENTRAL POSTEROMEDIAL NUCLEUS (pain & temperature) OF THE THALAMUS
(Synapse to 3rd order neuron)

TRIGEMINOTHALAMIC TRACT/TRIGEMINAL LEMNISCUS
(Carries discriminative touch/pressure & proprioception fibres)
& VENTRAL SPINAL TRIGEMINAL LEMNISCUS
(Carries pain & temperature fibres)

TRIGEMINAL SENSORY NUCLEI PRINCIPAL SENSORY NUCLEUS,
MESENCEPHALIC NUCLEUS, SPINAL NUCLEUS
(Responsible for touch & pressure, proprioception and pain & temperature
respectively; Synapse to 2nd order neuron)

TRIGEMINAL NERVE SENSORY ROOT

TRIGEMINAL GANGLION (SEMILUNAR GANGLIONS)

| OPTHALMIC NERVE | MAXILLARY NERVE | MANDIBULAR NERVE |
| (1st order neuron) | (1st order neuron) | (1st order neuron) |

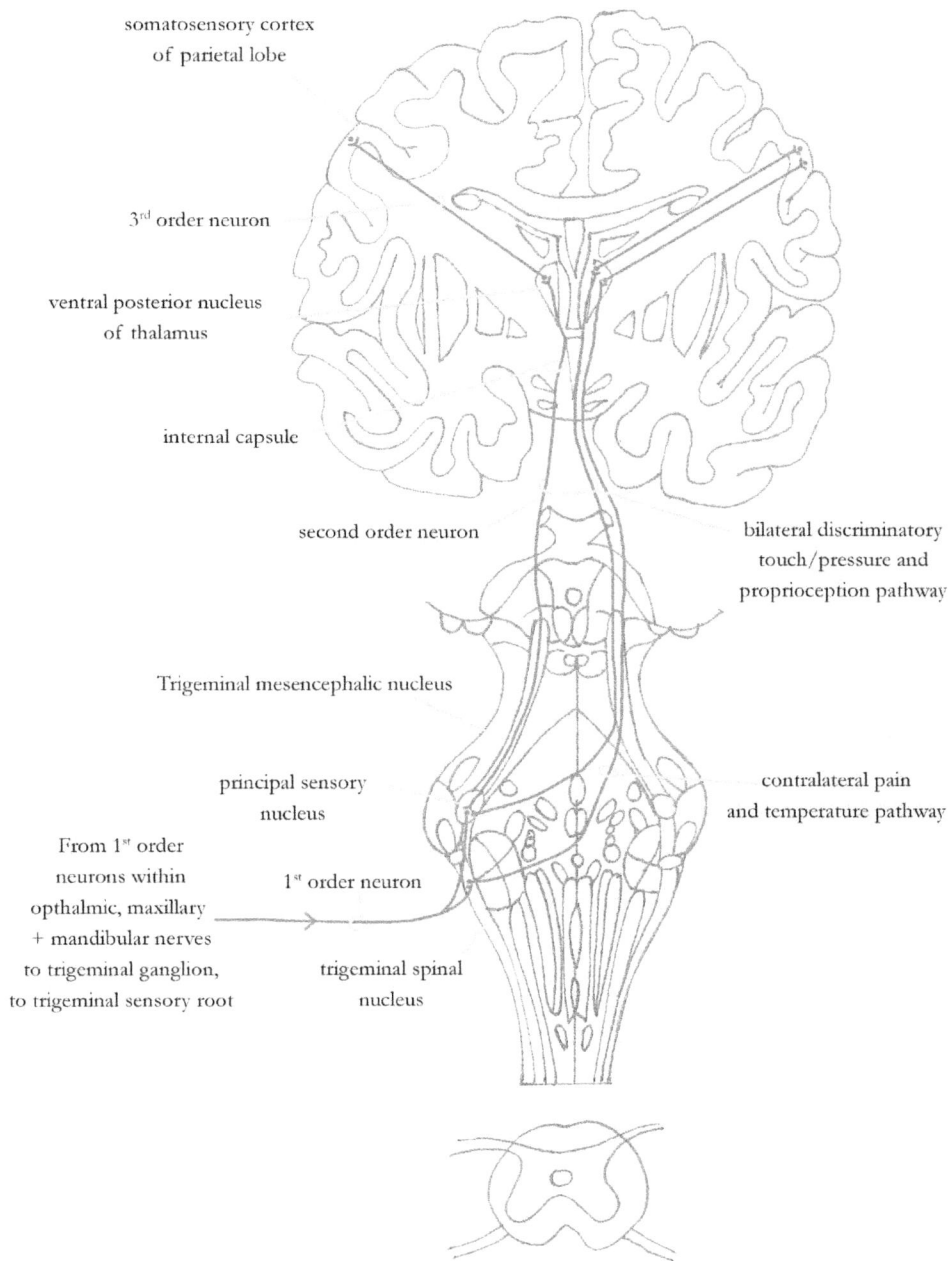

somatosensory cortex
of parietal lobe

3rd order neuron

ventral posterior nucleus
of thalamus

internal capsule

second order neuron

bilateral discriminatory
touch/pressure and
proprioception pathway

Trigeminal mesencephalic nucleus

principal sensory
nucleus

contralateral pain
and temperature pathway

From 1st order
neurons within
opthalmic, maxillary
+ mandibular nerves
to trigeminal ganglion,
to trigeminal sensory root

1st order neuron

trigeminal spinal
nucleus

Abducens Nerve CNVI

Somatic motor nerve to lateral rectus muscle of eye;
General somatic efferent

FIBRES PROJECT FROM THE CORTEX (FRONTAL EYE FIELDS, SUPPLEMENTARY EYE FIELDS & POSTERIOR PARIETAL CORTEX) CAUDATE NUCLEUS, SUBSTANTIA NIGRA, MEDIAL LONGITUDINAL FASCICULUS (CONNECTION TO OCULOMOTOR, TROCHLEAR & VESTIBULAR NUCLEI),TECTOBULBAR TRACT (FROM DEEP LAYERS OF SUPERIOR COLLICULUS), PARAMEDIAN PONTINE RETICULAR FORMATION
(Upper motor neuron)

ABDUCENS NUCLEUS IN PONS
(Fibres decussate to contralateral side proximal to abducens nucleus; synapse to lower motor neuron)

ABDUCENS NERVE
(Emerges through the ventral surface of brainstem at the junction of pyramid of the medulla and the pons)

PASSES THROUGH CAVERNOUS SINUS

EMERGES THROUGH SUPERIOR ORBITAL FISSURE

NEUROMUSCULAR JUNCTION OF LATERAL RECTUS MUSCLE OF EYE

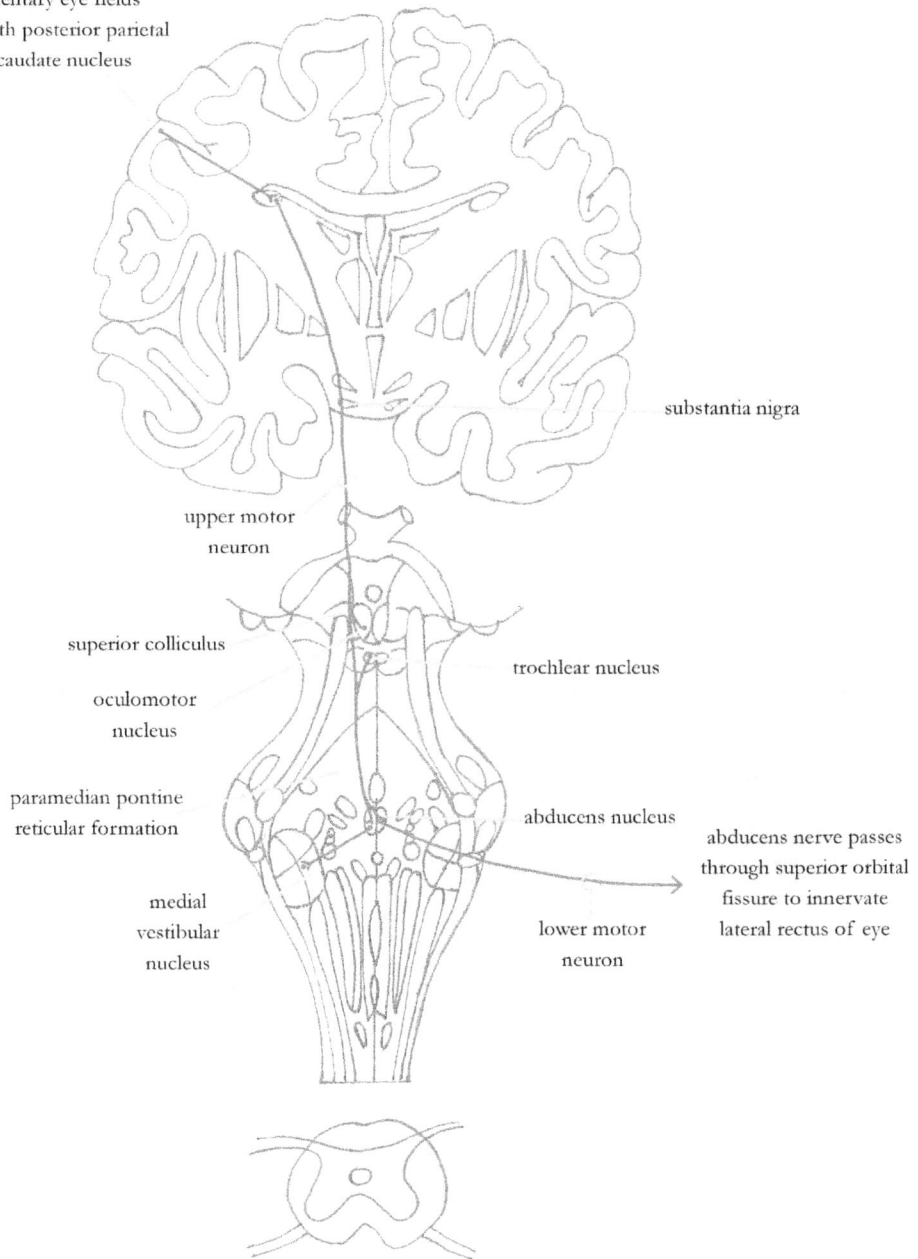

from cortex (frontal eye fields)
+ supplementary eye fields
connecting with posterior parietal
cortex + caudate nucleus

substantia nigra

upper motor
neuron

superior colliculus

trochlear nucleus

oculomotor
nucleus

paramedian pontine
reticular formation

abducens nucleus

abducens nerve passes
through superior orbital
fissure to innervate
lateral rectus of eye

medial
vestibular
nucleus

lower motor
neuron

* This image shows some of the connections during saccadic eye movement, smooth pursuit, gaze, active visual fixation and vergence.

Facial Nerve CNVII Sensory Pathway No. 1

General somatic afferent sensation to nasal and sinus cavities, velum, pinna of ear and external acoustic meatus

PRIMARY SENSORY/SOMATOSENSORY CORTEX OF CEREBRUM
(Brodmann Areas 1,2,3)

VENTRAL POSTERIOR NUCLEUS OF THE THALAMUS
(Synapse to 3rd order neuron)

SPINAL TRIGEMINAL NUCLEUS
(Synapse to 2nd order neuron; fibres cross to contralateral side & ascend through the trigeminal lemniscus to the thalamus; some fibres remain ipsilateral also)

GENICULATE GANGLION
(Housed within the facial canal of the petrous temporal bone)

FACIAL NERVE
(1st order neuron)

FIVE BRANCHES OF FACIAL NERVE
(TEMPORAL, ZYGOMATIC, BUCCAL, MANDIBULAR, CERVICAL)

NASAL CAVITY, SINUS CAVITIES, VELUM, PINNA OF THE EAR, EXTERNAL ACOUSTIC MEATUS

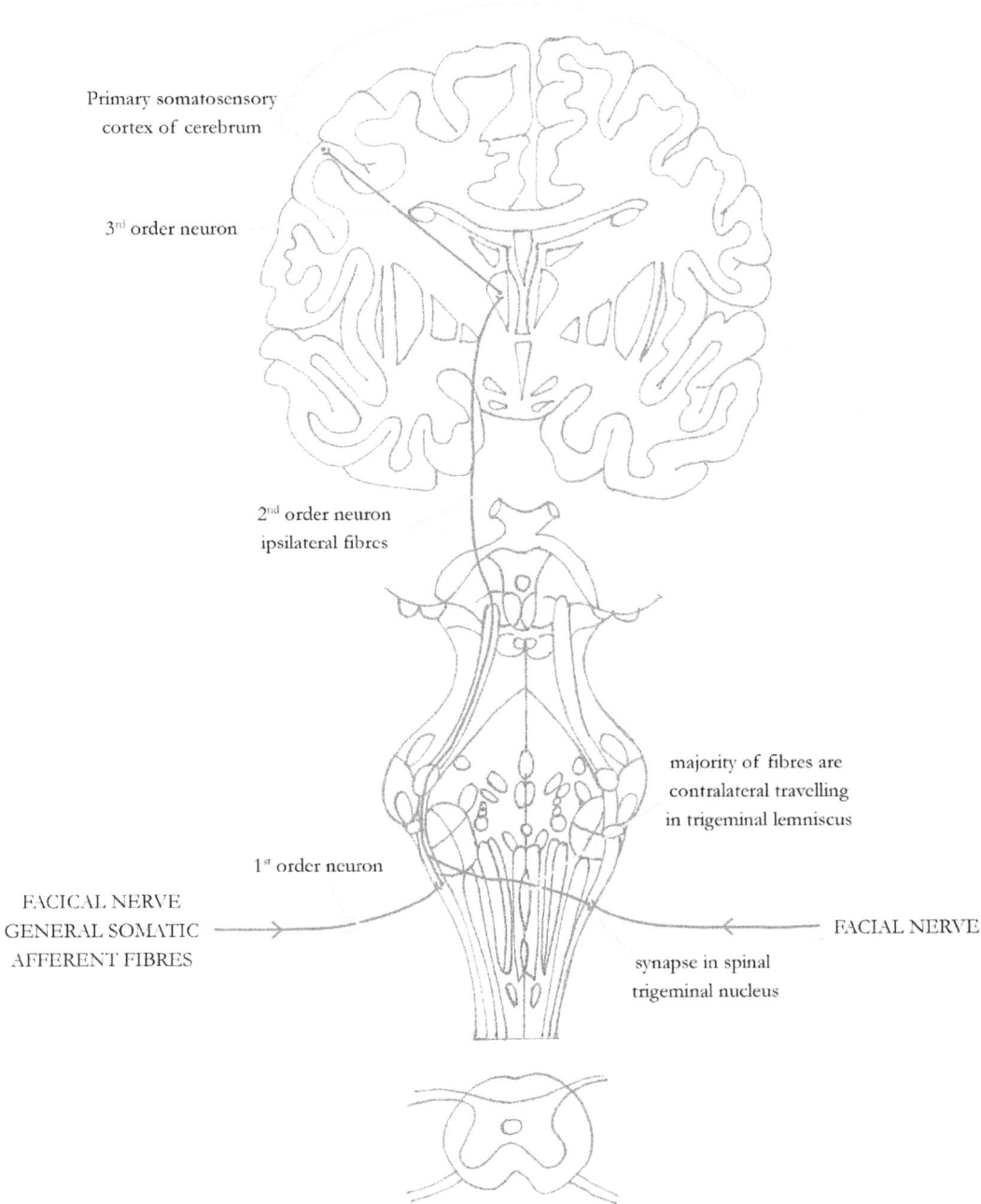

Primary somatosensory
cortex of cerebrum

3rd order neuron

2nd order neuron
ipsilateral fibres

majority of fibres are
contralateral travelling
in trigeminal lemniscus

1st order neuron

FACICAL NERVE
GENERAL SOMATIC
AFFERENT FIBRES

FACIAL NERVE

synapse in spinal
trigeminal nucleus

Facial Nerve CNVII Sensory Pathway No. 2

Special visceral afferent sensation of taste to anterior $^2/_3$ of tongue

FRONTAL OPERCULUM, ANTERIOR INSULAR CORTEX
AND ROSTRAL PART OF BRODMANN AREA 3B

PASSES THROUGH POSTERIOR LIMB OF INTERNAL CAPSULE

VENTRAL POSTERIOR NUCLEUS OF THE THALAMUS
(Synapse to 3rd order neuron)

ROSTRAL PART OF NUCLEUS SOLITARIUS GUSTATORY REGION
WITHIN MEDULLA OBLONGATA
(Synapse to 2nd order neuron)

GENICULATE GANGLION

FACIAL NERVE SENSORY ROOT
(1st order neuron)

CHORDA TYMPANI NERVE   GREATER PETROSAL NERVE

TASTE RECEPTORS OF   TASTE RECEPTORS OF VELUM
ANTERIOR 2/3 OF TONGUE

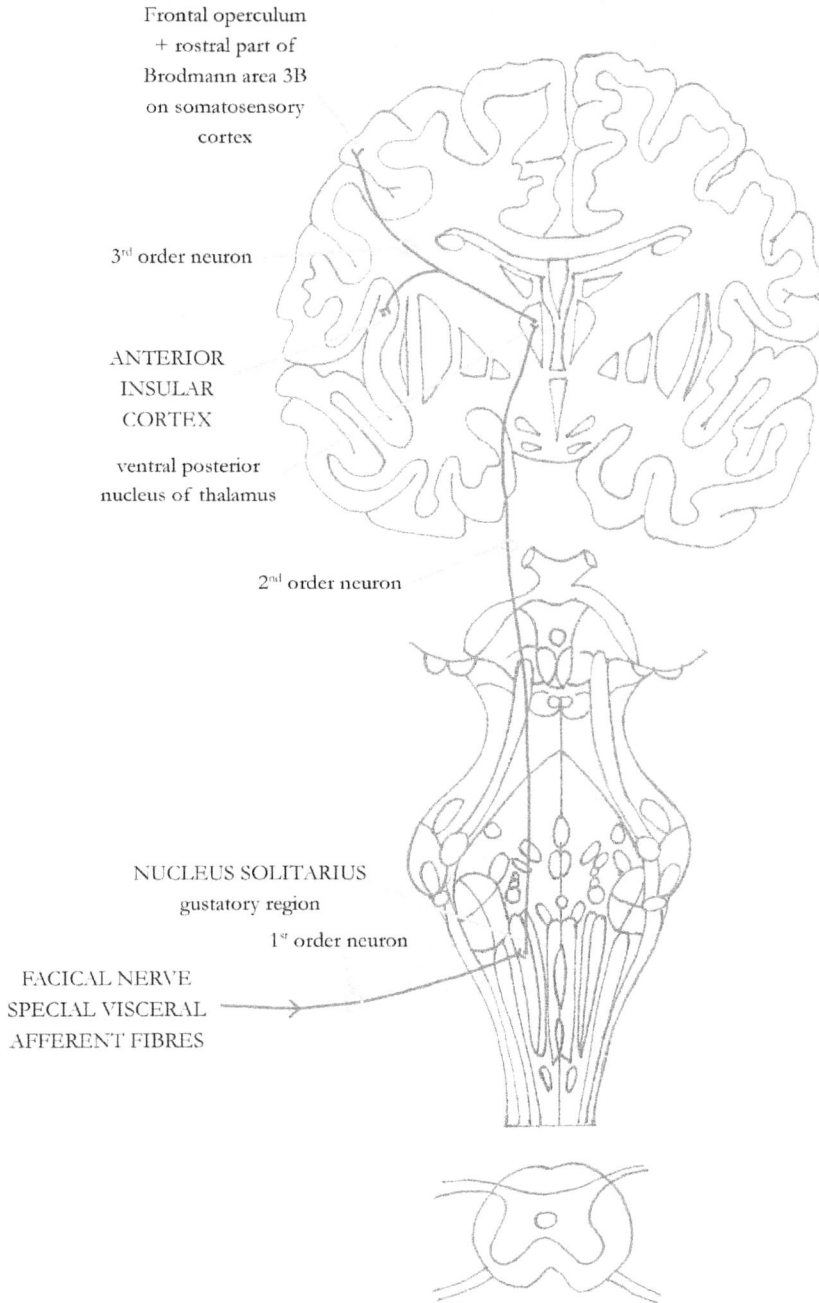

Frontal operculum
+ rostral part of
Brodmann area 3B
on somatosensory
cortex

3rd order neuron

ANTERIOR
INSULAR
CORTEX

ventral posterior
nucleus of thalamus

2nd order neuron

NUCLEUS SOLITARIUS
gustatory region

1st order neuron

FACICAL NERVE
SPECIAL VISCERAL
AFFERENT FIBRES

Facial Nerve CNVII Branchial Motor Pathway

Special visceral efferent innervation of all muscles of facial expression, posterior belly of digastric, stylohyoid, platysma, stapedius, superior & inferior auricular & occipitofrontalis muscles

PRIMARY MOTOR CORTEX OF CEREBRUM
(Upper motor neuron)

CORTICOBULBAR FIBRES
(Descending through the corona radiata and genu of the internal capsule)

FACIAL NUCLEUS WITHIN CAUDAL PONTINE TEGMENTUM
(Synapse to lower motor neuron; upper face & scalp receives innervation from both ipsilateral & contralateral corticobulbar fibres, lower face receives only contralateral innervation)

MOTOR ROOT OF THE FACIAL NERVE
(Emerges from the pontomedullary junction)

FACIAL NERVE
(Temporal, zygomatic, buccal, mandibular and cervical branches)

NEUROMUSCULAR JUNCTION OF FACIAL MUSCLES
(All muscles of facial expression, posterior belly of digastric, stylohyoid, platysma, stapedius, superior & inferior auricular & occipitofrontalis muscles)

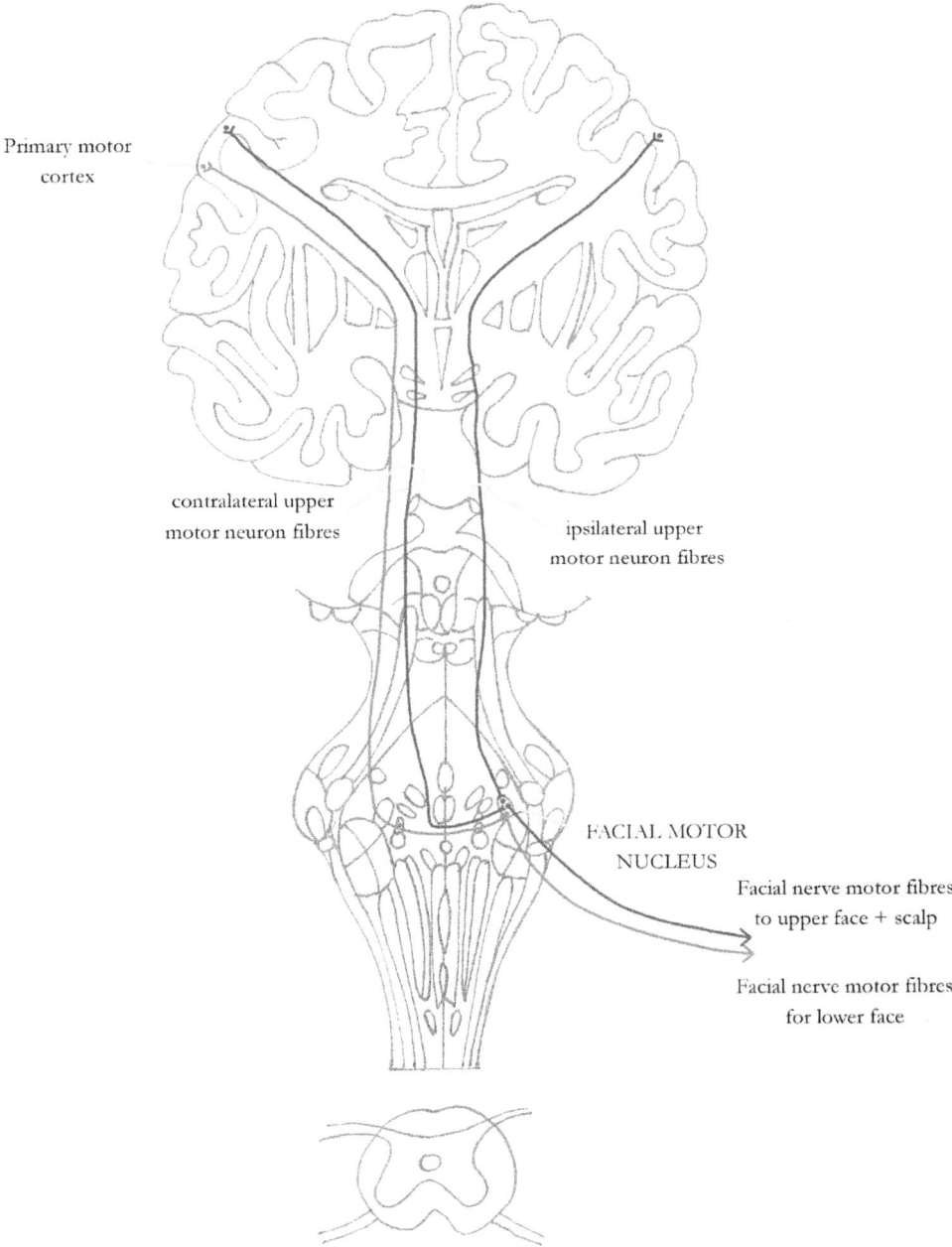

Primary motor
cortex

contralateral upper
motor neuron fibres

ipsilateral upper
motor neuron fibres

FACIAL MOTOR
NUCLEUS

Facial nerve motor fibres
to upper face + scalp

Facial nerve motor fibres
for lower face

Facial Nerve CNVII Efferent Parasympathetic Pathway

General visceral efferent innervation to nasal & oral cavity, soft & hard palate mucosa, palatine, lacrimal, nasal, submandibular & sublingual glands

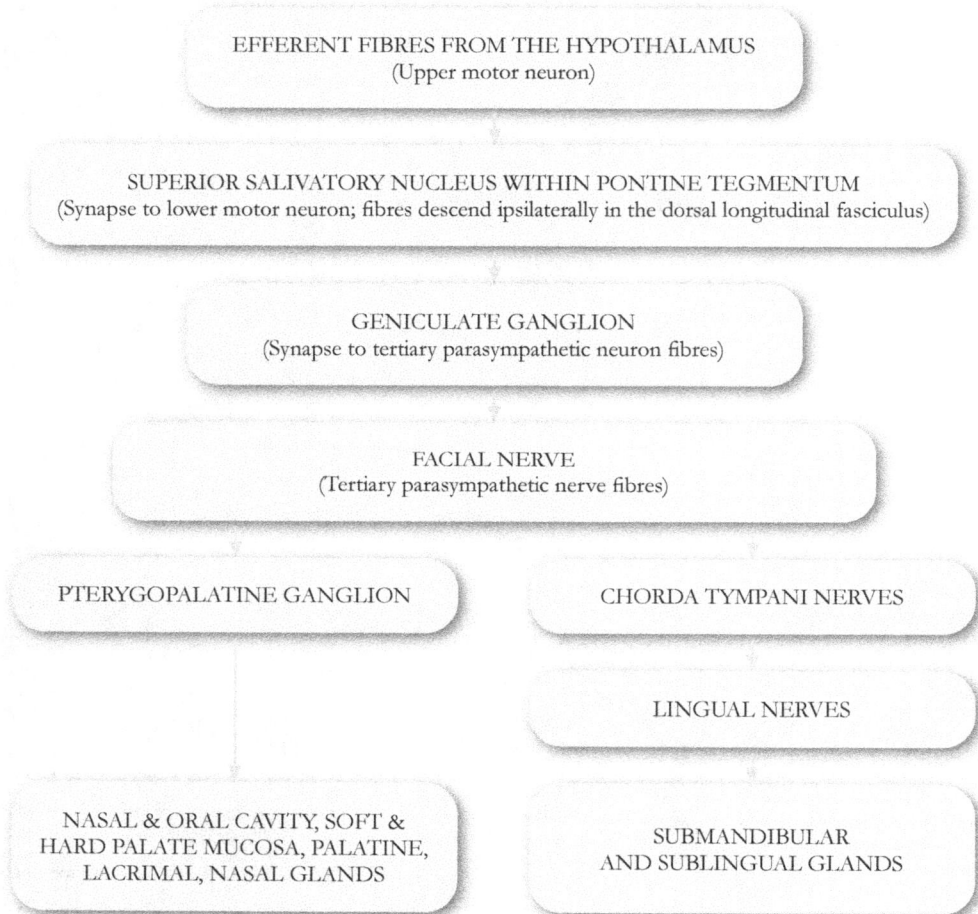

EFFERENT FIBRES FROM THE HYPOTHALAMUS
(Upper motor neuron)

SUPERIOR SALIVATORY NUCLEUS WITHIN PONTINE TEGMENTUM
(Synapse to lower motor neuron; fibres descend ipsilaterally in the dorsal longitudinal fasciculus)

GENICULATE GANGLION
(Synapse to tertiary parasympathetic neuron fibres)

FACIAL NERVE
(Tertiary parasympathetic nerve fibres)

PTERYGOPALATINE GANGLION

CHORDA TYMPANI NERVES

LINGUAL NERVES

NASAL & ORAL CAVITY, SOFT & HARD PALATE MUCOSA, PALATINE, LACRIMAL, NASAL GLANDS

SUBMANDIBULAR AND SUBLINGUAL GLANDS

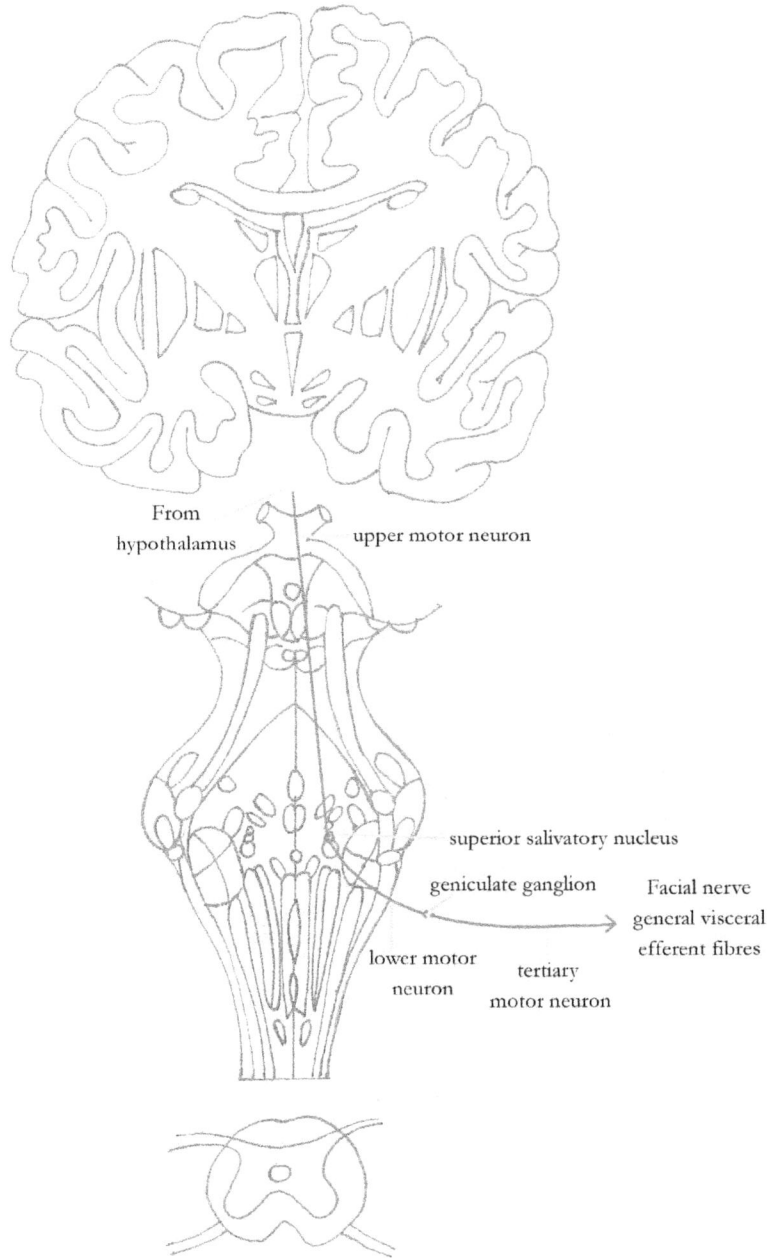

From
hypothalamus

upper motor neuron

superior salivatory nucleus

geniculate ganglion

Facial nerve
general visceral
efferent fibres

lower motor
neuron

tertiary
motor neuron

Vestibulocochlear Nerve CNVIII Special Somatic Afferent
1. Vestibular Nerve

Position and movement of head

PRIMARY SENSORY/SOMATOSENSORY CORTEX OF CEREBRUM
(Primary vestibular cortical area located in the parietal lobe at the junction between the intraparietal and post-central sulci, which is adjacent to the area where the head is represented; the area for conscious appreciation of body position)

VENTRAL POSTERIOR & MEDIAL PULVINAR NUCLEUS OF THALAMUS
(Fibres ascend to the contralateral thalamus; synapse to 3rd order neuron)

CEREBELLUM (VESTIBULOCEREBELLAR FIBRES)
(Fibres project to flocculonodular lobe; fibres travel ipsilateral)

ABDUCENS, TROCHLEAR & OCULOMOTOR NERVES
(For coordination of head & eye movements; fibres travel ipsilateral)

MEDIAL LONGITUDINAL FASCICULUS

VESTIBULAR NUCLEI IN ROSTRAL MEDULLA OBLONGATA
(Superior, inferior, medial & lateral vestibular nuclei located beneath the floor of the 4th ventricle; synapse to 2nd order neuron)

VESTIBULAR (SCARPA) GANGLION
(Located within the internal auditory meatus)

VESTIBULAR NERVE (1st order neuron)
DENDRITES ON HAIR CELLS OF VESTIBULAR PORTION
OF MEMBRANOUS LABYRINTH OF SEMICIRCULAR CANALS

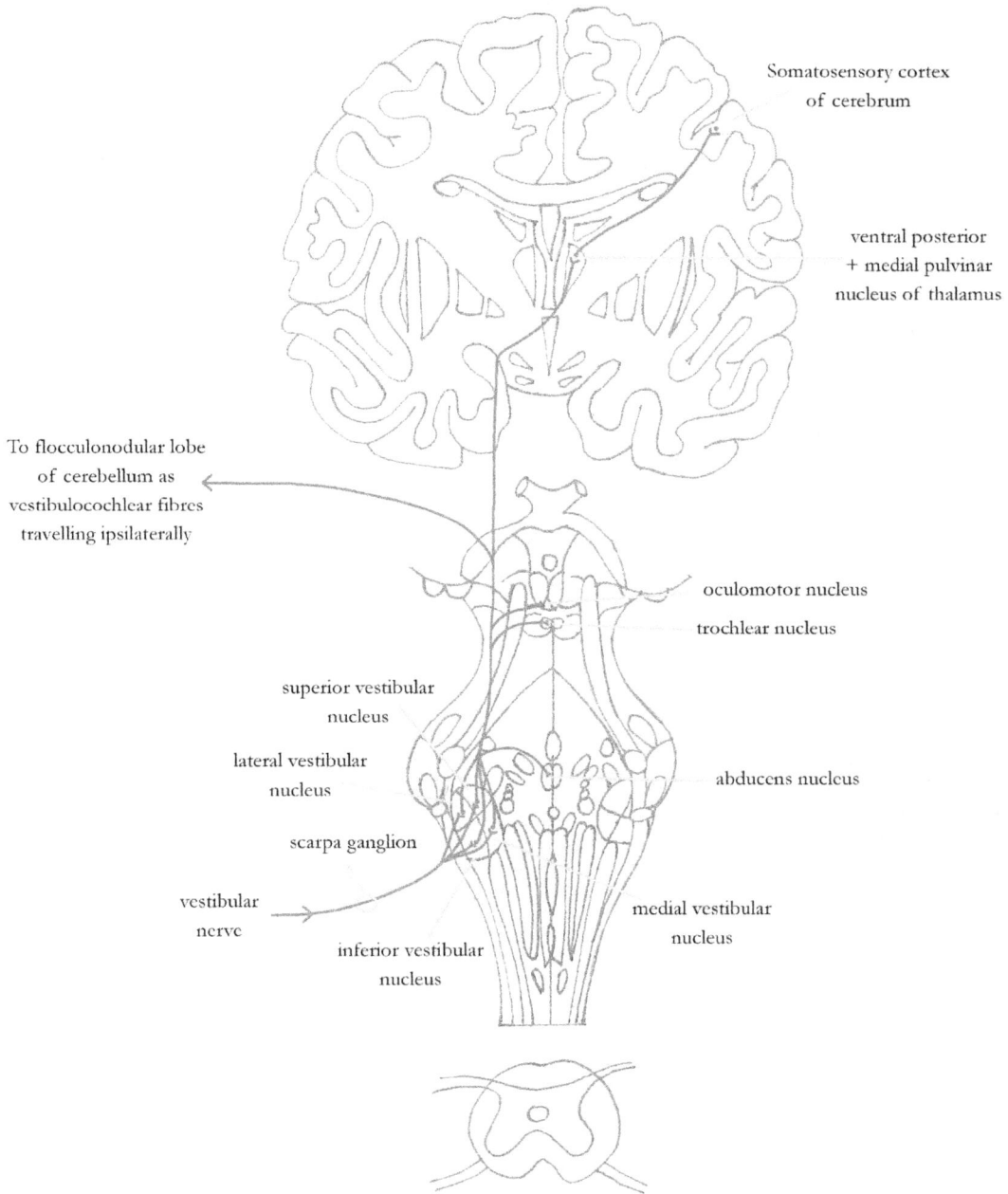

Somatosensory cortex
of cerebrum

ventral posterior
+ medial pulvinar
nucleus of thalamus

To flocculonodular lobe
of cerebellum as
vestibulocochlear fibres
travelling ipsilaterally

oculomotor nucleus

trochlear nucleus

superior vestibular
nucleus

lateral vestibular
nucleus

abducens nucleus

scarpa ganglion

vestibular
nerve

medial vestibular
nucleus

inferior vestibular
nucleus

Vestibulocochlear Nerve CNVIII Special Somatic Afferent
2. Cochlear Nerve
The sense of hearing

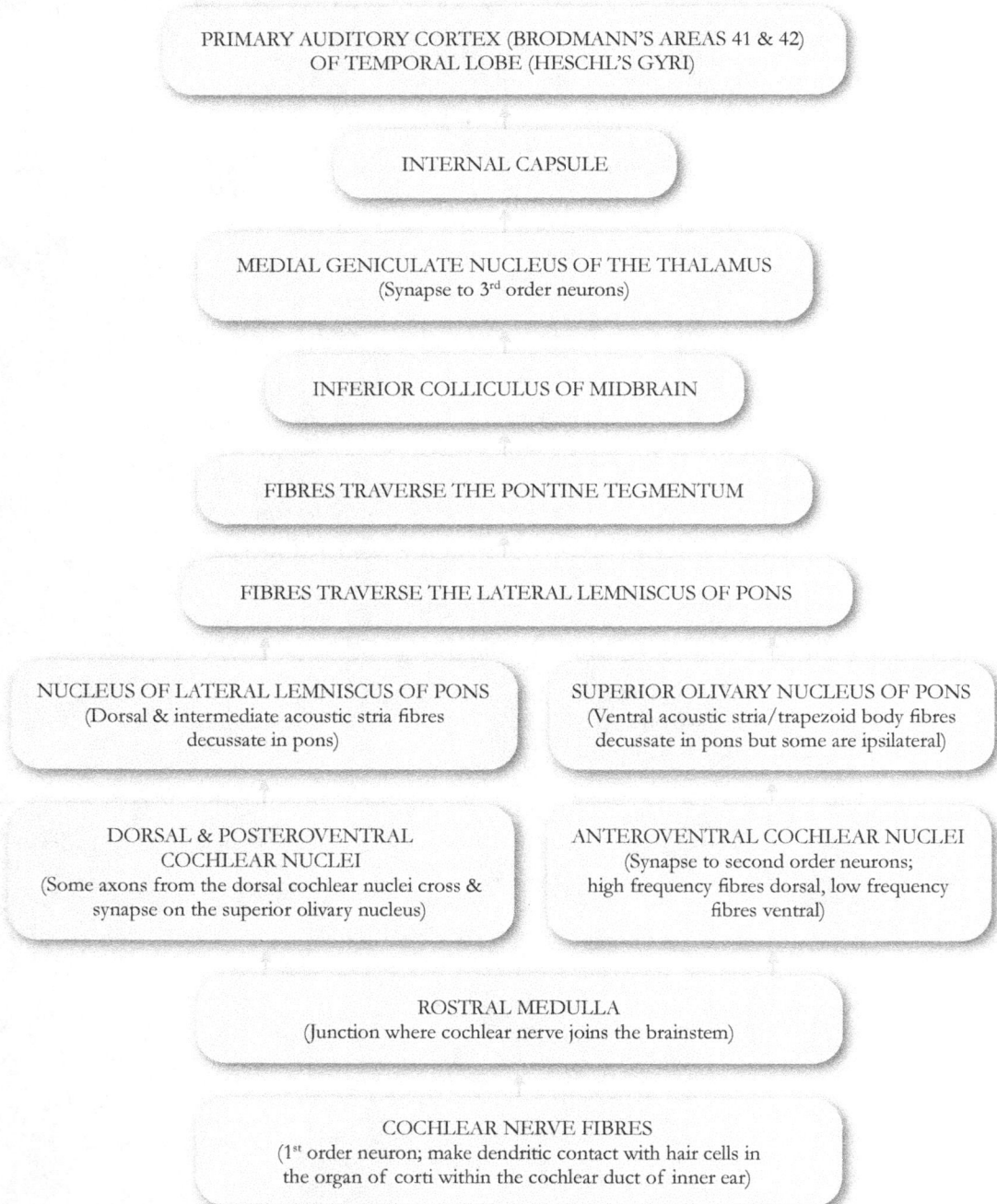

PRIMARY AUDITORY CORTEX (BRODMANN'S AREAS 41 & 42) OF TEMPORAL LOBE (HESCHL'S GYRI)

INTERNAL CAPSULE

MEDIAL GENICULATE NUCLEUS OF THE THALAMUS
(Synapse to 3rd order neurons)

INFERIOR COLLICULUS OF MIDBRAIN

FIBRES TRAVERSE THE PONTINE TEGMENTUM

FIBRES TRAVERSE THE LATERAL LEMNISCUS OF PONS

NUCLEUS OF LATERAL LEMNISCUS OF PONS
(Dorsal & intermediate acoustic stria fibres decussate in pons)

SUPERIOR OLIVARY NUCLEUS OF PONS
(Ventral acoustic stria/trapezoid body fibres decussate in pons but some are ipsilateral)

DORSAL & POSTEROVENTRAL COCHLEAR NUCLEI
(Some axons from the dorsal cochlear nuclei cross & synapse on the superior olivary nucleus)

ANTEROVENTRAL COCHLEAR NUCLEI
(Synapse to second order neurons; high frequency fibres dorsal, low frequency fibres ventral)

ROSTRAL MEDULLA
(Junction where cochlear nerve joins the brainstem)

COCHLEAR NERVE FIBRES
(1st order neuron; make dendritic contact with hair cells in the organ of corti within the cochlear duct of inner ear)

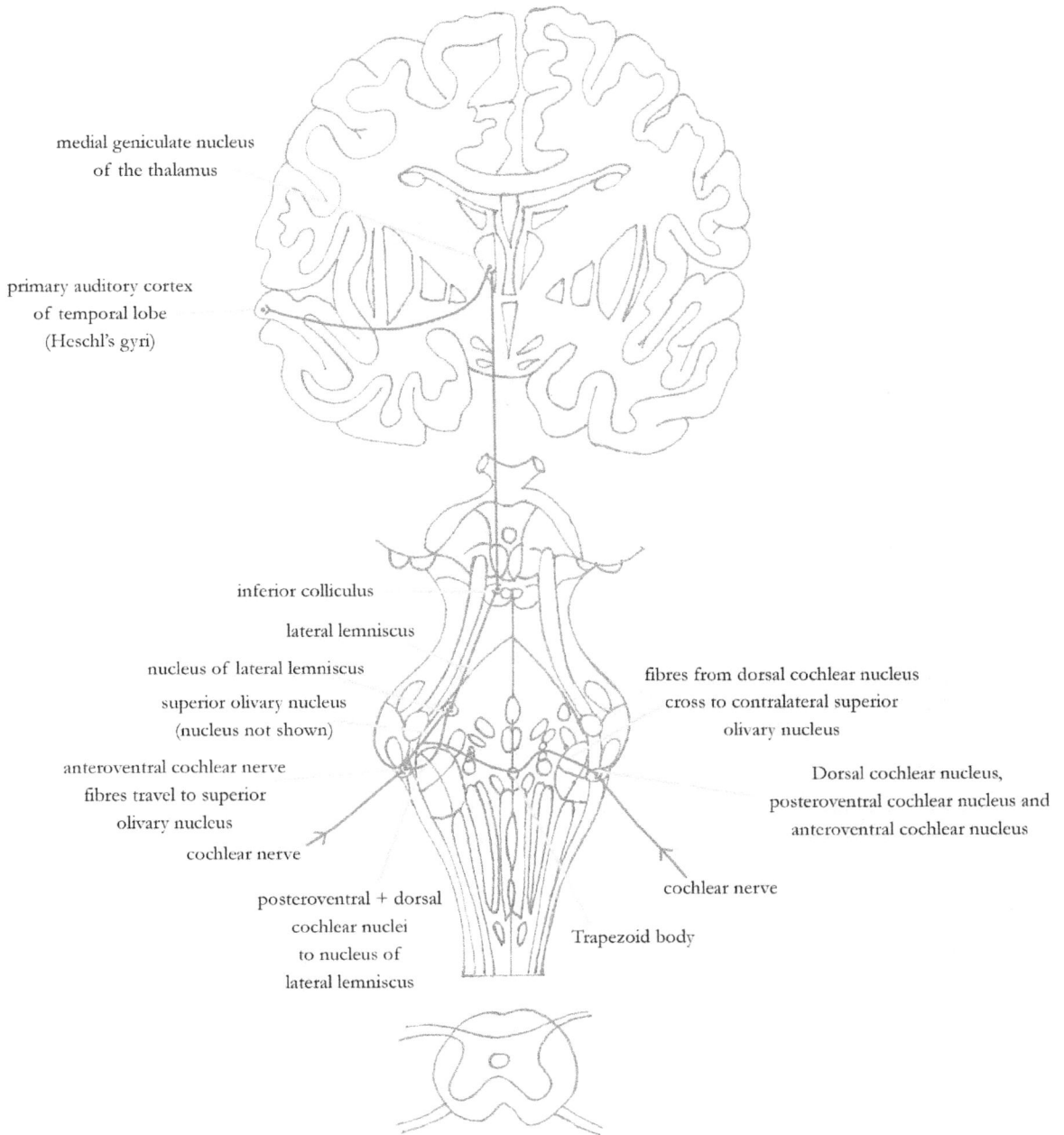

medial geniculate nucleus
of the thalamus

primary auditory cortex
of temporal lobe
(Heschl's gyri)

inferior colliculus

lateral lemniscus

nucleus of lateral lemniscus

superior olivary nucleus
(nucleus not shown)

anteroventral cochlear nerve
fibres travel to superior
olivary nucleus

cochlear nerve

posteroventral + dorsal
cochlear nuclei
to nucleus of
lateral lemniscus

fibres from dorsal cochlear nucleus
cross to contralateral superior
olivary nucleus

Dorsal cochlear nucleus,
posteroventral cochlear nucleus and
anteroventral cochlear nucleus

cochlear nerve

Trapezoid body

Glossopharyngeal Nerve CNIX General Somatic Afferent

*General sensation of pharyngeal tonsil, posterior $1/3$ of tongue
Eustachian tube and middle ear*

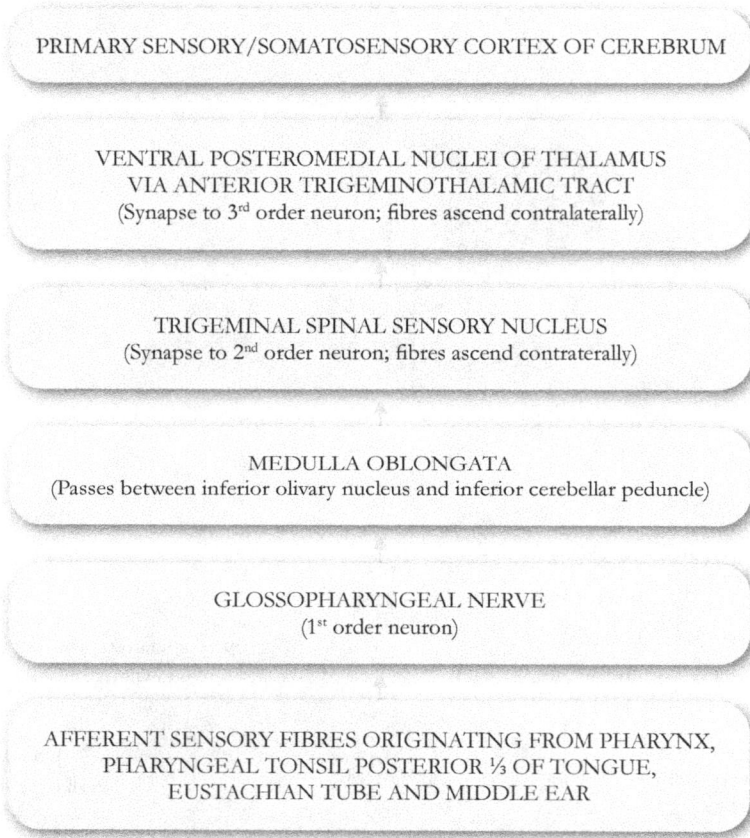

PRIMARY SENSORY/SOMATOSENSORY CORTEX OF CEREBRUM

VENTRAL POSTEROMEDIAL NUCLEI OF THALAMUS
VIA ANTERIOR TRIGEMINOTHALAMIC TRACT
(Synapse to 3rd order neuron; fibres ascend contralaterally)

TRIGEMINAL SPINAL SENSORY NUCLEUS
(Synapse to 2nd order neuron; fibres ascend contraterally)

MEDULLA OBLONGATA
(Passes between inferior olivary nucleus and inferior cerebellar peduncle)

GLOSSOPHARYNGEAL NERVE
(1st order neuron)

AFFERENT SENSORY FIBRES ORIGINATING FROM PHARYNX,
PHARYNGEAL TONSIL POSTERIOR ⅓ OF TONGUE,
EUSTACHIAN TUBE AND MIDDLE EAR

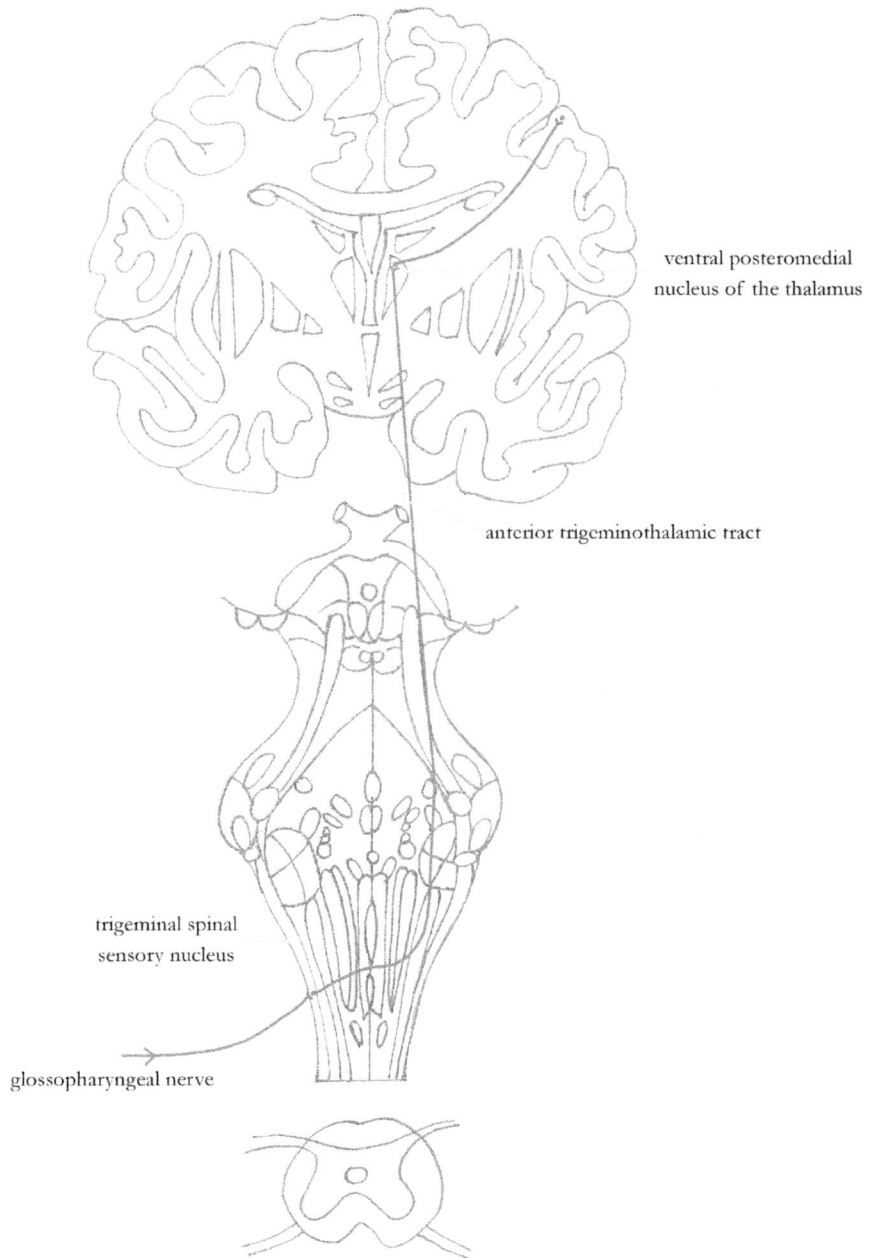

ventral posteromedial
nucleus of the thalamus

anterior trigeminothalamic tract

trigeminal spinal
sensory nucleus

glossopharyngeal nerve

Glossopharyngeal Nerve CNIX Special Visceral Afferent

Taste receptors on posterior $^1/_3$ of tongue

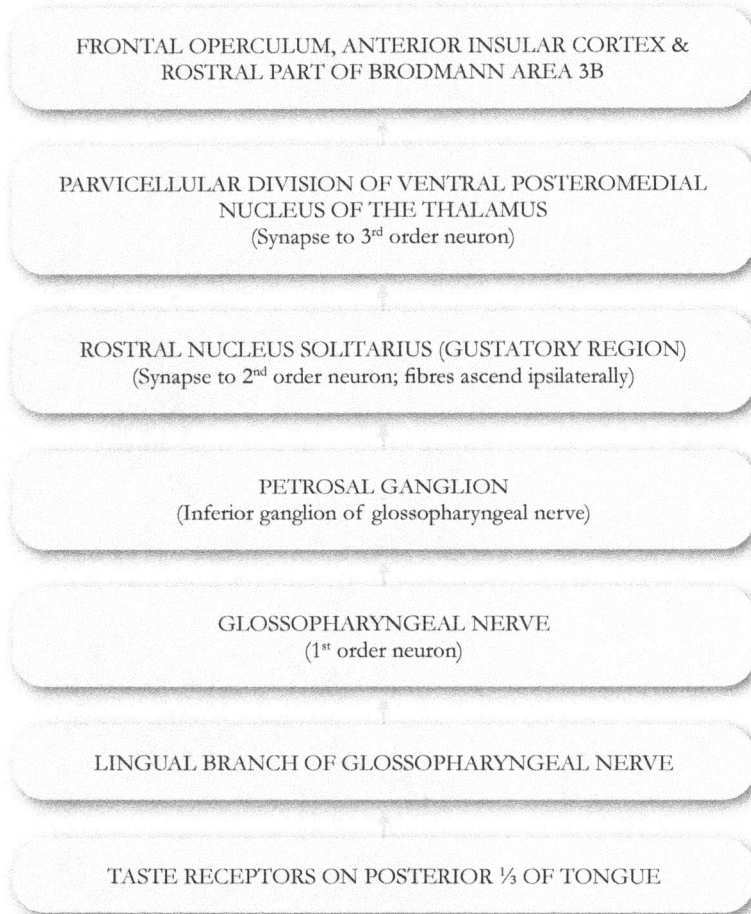

FRONTAL OPERCULUM, ANTERIOR INSULAR CORTEX &
ROSTRAL PART OF BRODMANN AREA 3B

PARVICELLULAR DIVISION OF VENTRAL POSTEROMEDIAL
NUCLEUS OF THE THALAMUS
(Synapse to 3rd order neuron)

ROSTRAL NUCLEUS SOLITARIUS (GUSTATORY REGION)
(Synapse to 2nd order neuron; fibres ascend ipsilaterally)

PETROSAL GANGLION
(Inferior ganglion of glossopharyngeal nerve)

GLOSSOPHARYNGEAL NERVE
(1st order neuron)

LINGUAL BRANCH OF GLOSSOPHARYNGEAL NERVE

TASTE RECEPTORS ON POSTERIOR ⅓ OF TONGUE

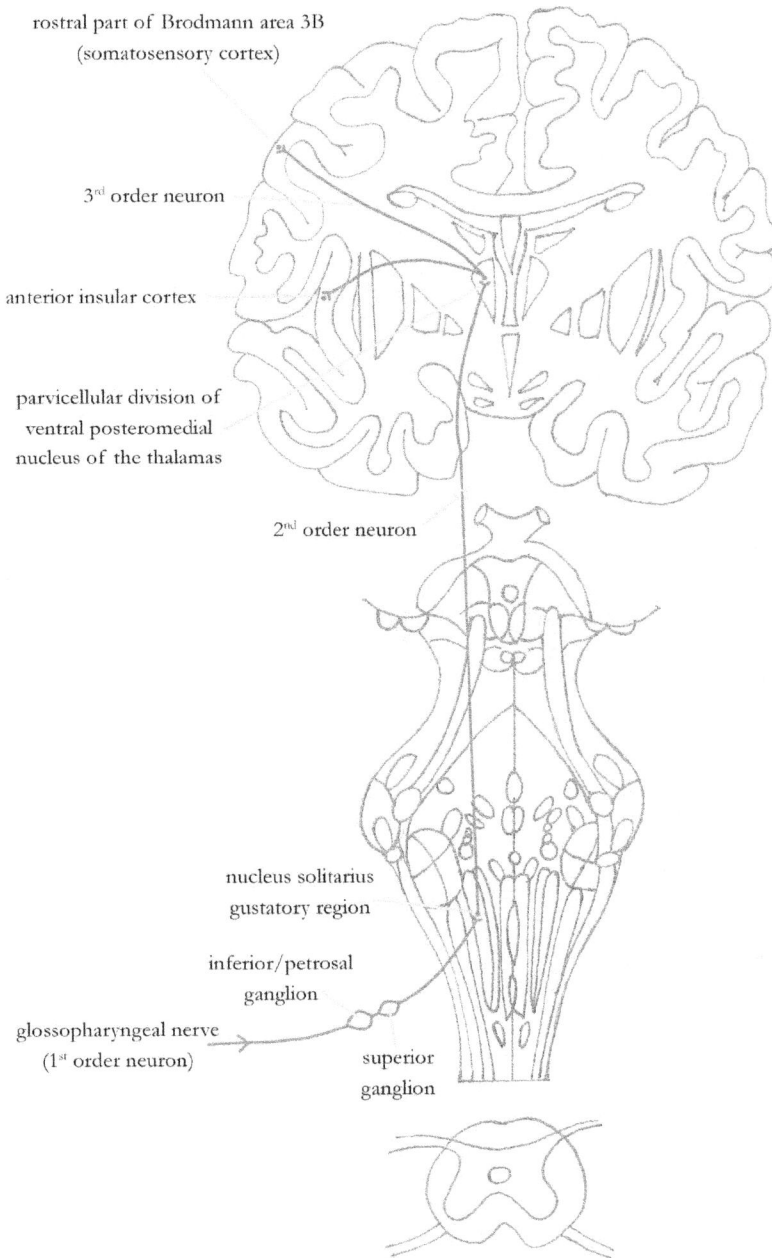

rostral part of Brodmann area 3B
(somatosensory cortex)

3rd order neuron

anterior insular cortex

parvicellular division of
ventral posteromedial
nucleus of the thalamas

2nd order neuron

nucleus solitarius
gustatory region

inferior/petrosal
ganglion

glossopharyngeal nerve
(1st order neuron)

superior
ganglion

Glossopharyngeal Nerve CNIX Branchial Motor Pathway

Special visceral efferent to stylopharyngeus muscle

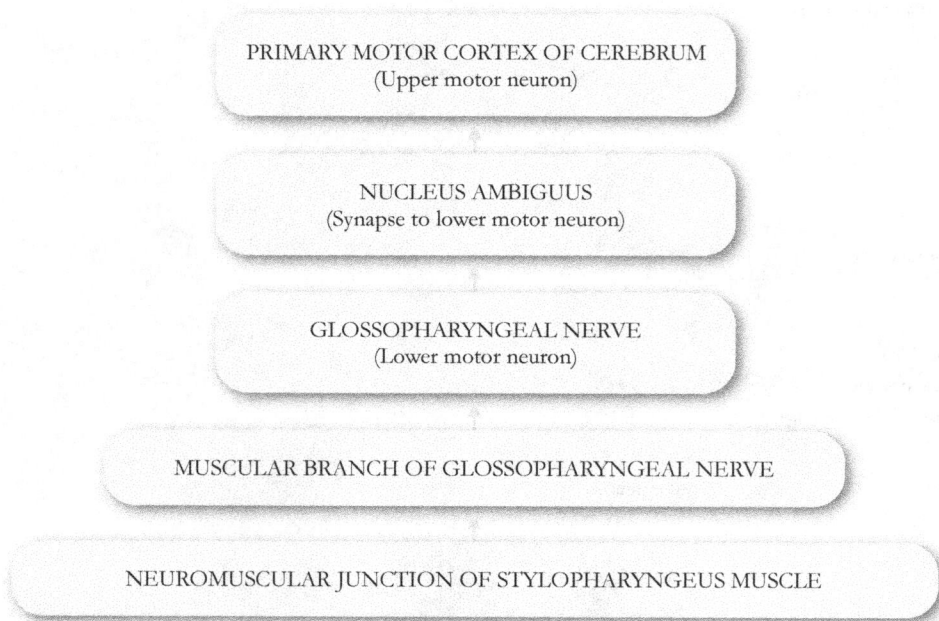

PRIMARY MOTOR CORTEX OF CEREBRUM
(Upper motor neuron)

NUCLEUS AMBIGUUS
(Synapse to lower motor neuron)

GLOSSOPHARYNGEAL NERVE
(Lower motor neuron)

MUSCULAR BRANCH OF GLOSSOPHARYNGEAL NERVE

NEUROMUSCULAR JUNCTION OF STYLOPHARYNGEUS MUSCLE

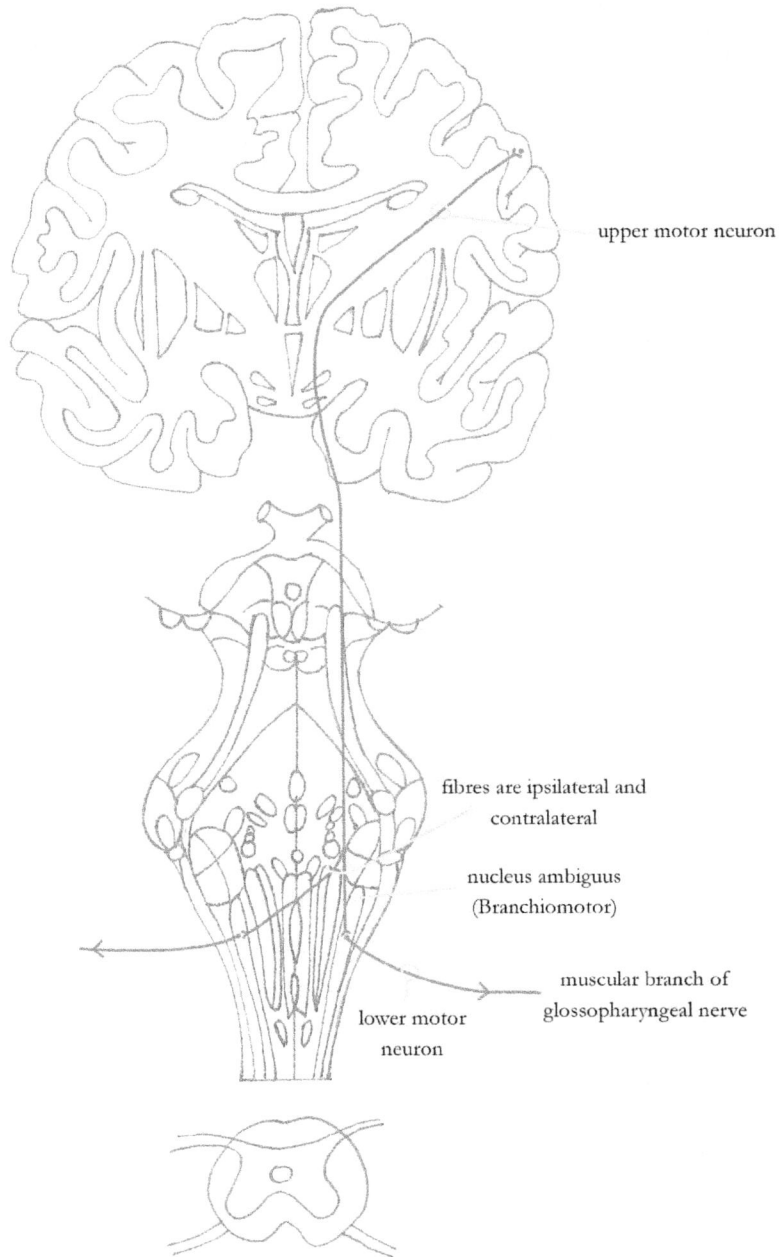

upper motor neuron

fibres are ipsilateral and
contralateral

nucleus ambiguus
(Branchiomotor)

muscular branch of
glossopharyngeal nerve

lower motor
neuron

Glossopharyngeal Nerve CNIX
Afferent Parasympathetic Pathway

To carotid sinus baroreceptors & carotid body chemoreceptors
General visceral afferent

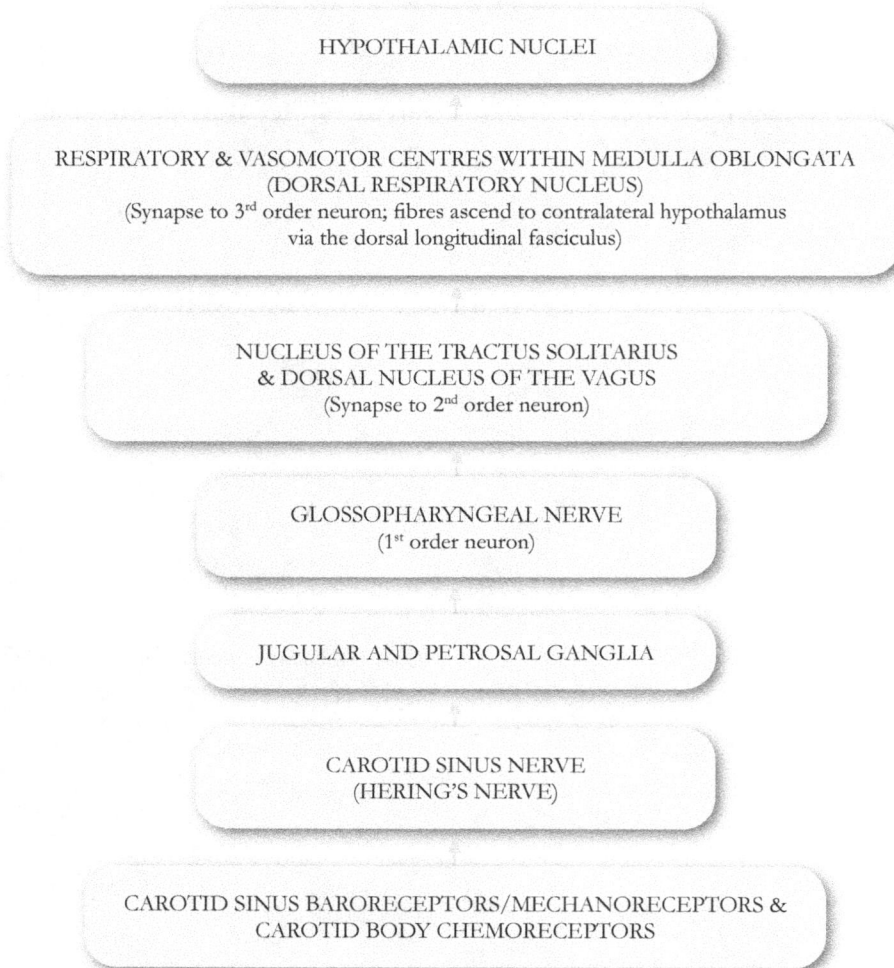

HYPOTHALAMIC NUCLEI

RESPIRATORY & VASOMOTOR CENTRES WITHIN MEDULLA OBLONGATA
(DORSAL RESPIRATORY NUCLEUS)
(Synapse to 3rd order neuron; fibres ascend to contralateral hypothalamus
via the dorsal longitudinal fasciculus)

NUCLEUS OF THE TRACTUS SOLITARIUS
& DORSAL NUCLEUS OF THE VAGUS
(Synapse to 2nd order neuron)

GLOSSOPHARYNGEAL NERVE
(1st order neuron)

JUGULAR AND PETROSAL GANGLIA

CAROTID SINUS NERVE
(HERING'S NERVE)

CAROTID SINUS BARORECEPTORS/MECHANORECEPTORS &
CAROTID BODY CHEMORECEPTORS

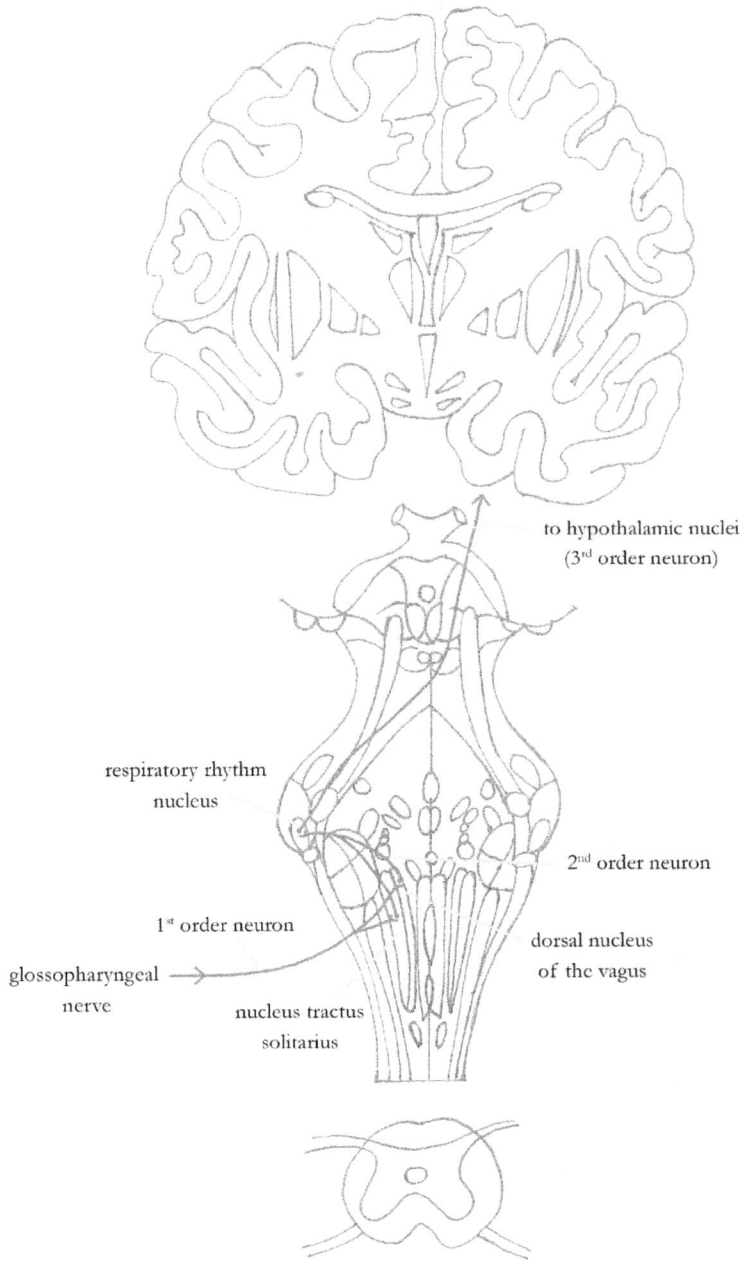

to hypothalamic nuclei
(3rd order neuron)

respiratory rhythm
nucleus

2nd order neuron

1st order neuron

dorsal nucleus
of the vagus

glossopharyngeal
nerve

nucleus tractus
solitarius

Glossopharyngeal Nerve CNIX
Efferent Parasympathetic Pathway

Innervation of parotid gland
General visceral efferent

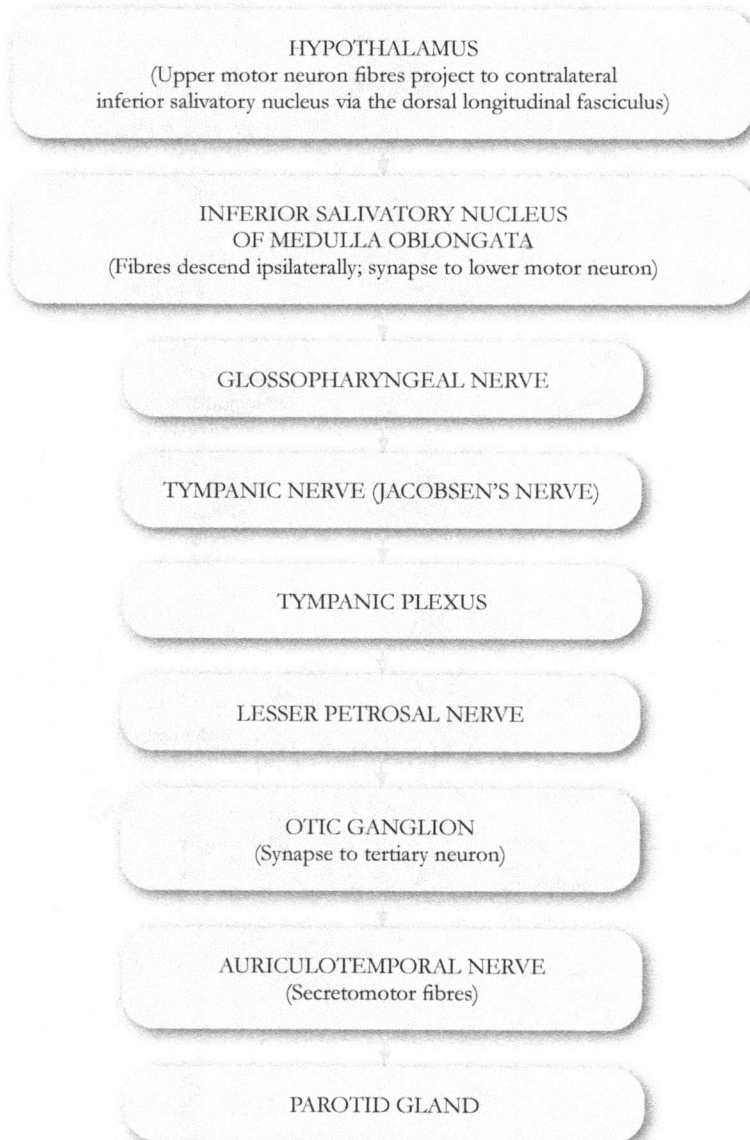

HYPOTHALAMUS
(Upper motor neuron fibres project to contralateral
inferior salivatory nucleus via the dorsal longitudinal fasciculus)

INFERIOR SALIVATORY NUCLEUS
OF MEDULLA OBLONGATA
(Fibres descend ipsilaterally; synapse to lower motor neuron)

GLOSSOPHARYNGEAL NERVE

TYMPANIC NERVE (JACOBSEN'S NERVE)

TYMPANIC PLEXUS

LESSER PETROSAL NERVE

OTIC GANGLION
(Synapse to tertiary neuron)

AURICULOTEMPORAL NERVE
(Secretomotor fibres)

PAROTID GLAND

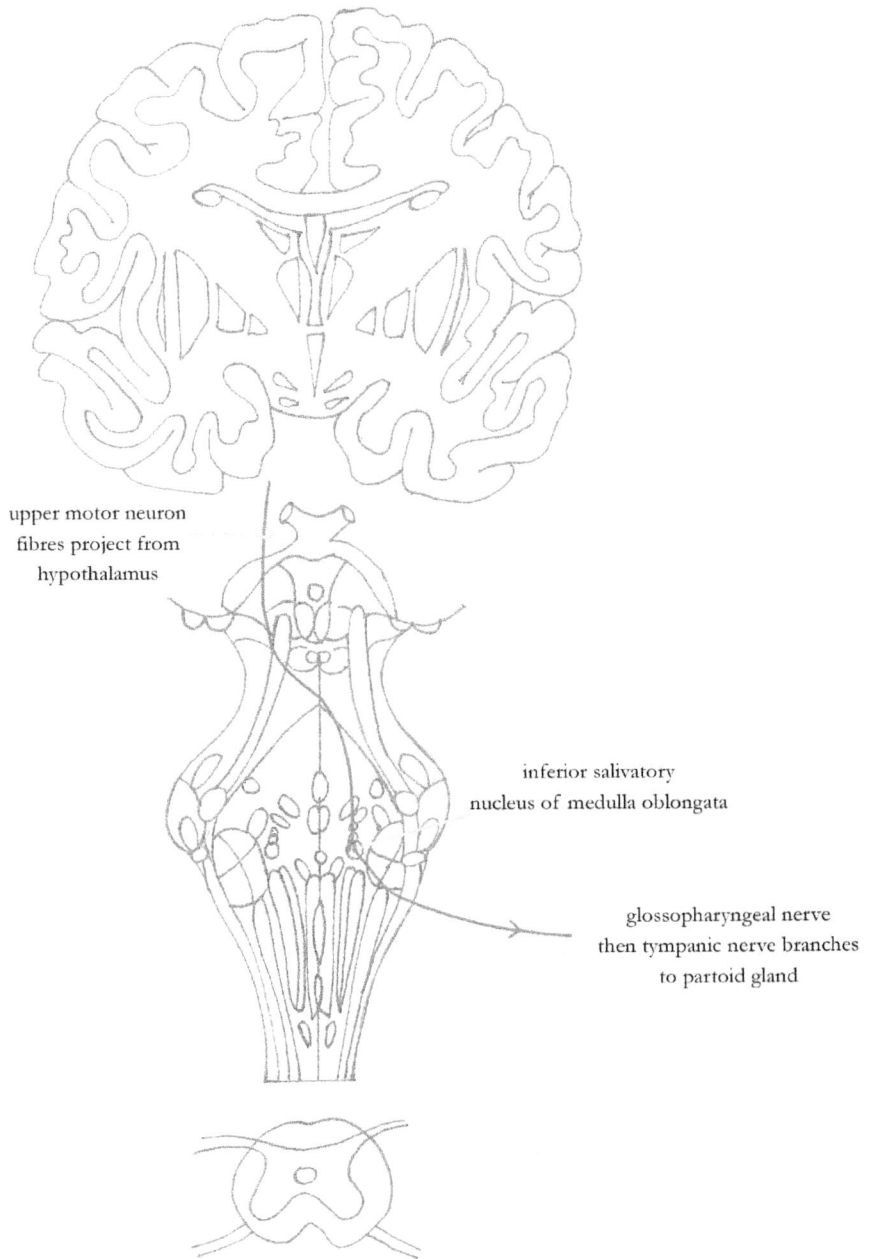

upper motor neuron
fibres project from
hypothalamus

inferior salivatory
nucleus of medulla oblongata

glossopharyngeal nerve
then tympanic nerve branches
to partoid gland

Vagus Nerve CNX Sensory Pathway

Somatic and visceral afferents
General somatic afferent & general visceral afferent

'SOMATOSENSORY CORTEX PROJECTIONS TO
HYPOTHALAMUS, AMYGDALA, & CEREBRAL INSULAR CORTEX
(VISCERAL SENSORY CORTEX)

VISCERAL SENSORY AREAS OF THALAMUS
INCLUDING VENTROPOSTERIOR PARVOCELLULAR NUCLEUS
(Synapse to 3rd order neuron)

TRAVERSES PONS & MIDBRAIN

TRIGEMINAL SENSORY NUCLEAS (general sensation)
& NUCLEUS SOLITARIUS (visceral sensation)
(Synapse to 2nd order neuron)

ROOTLETS OF VAGUS NERVE
(Lateral aspect of the medulla oblongata)

BRANCHES OF VAGUS NERVE
(1st order neuron)

RECEPTORS FOR GENERAL SENSATION OF PHARYNX,
LARYNX, OESOPHAGUS, TYMPANIC MEMBRANE, EXTERNAL
AUDITORY MEATUS, CONCHA OF EXTERNAL EAR THORACIC AND
ABDOMINAL VISCERA SENSATION

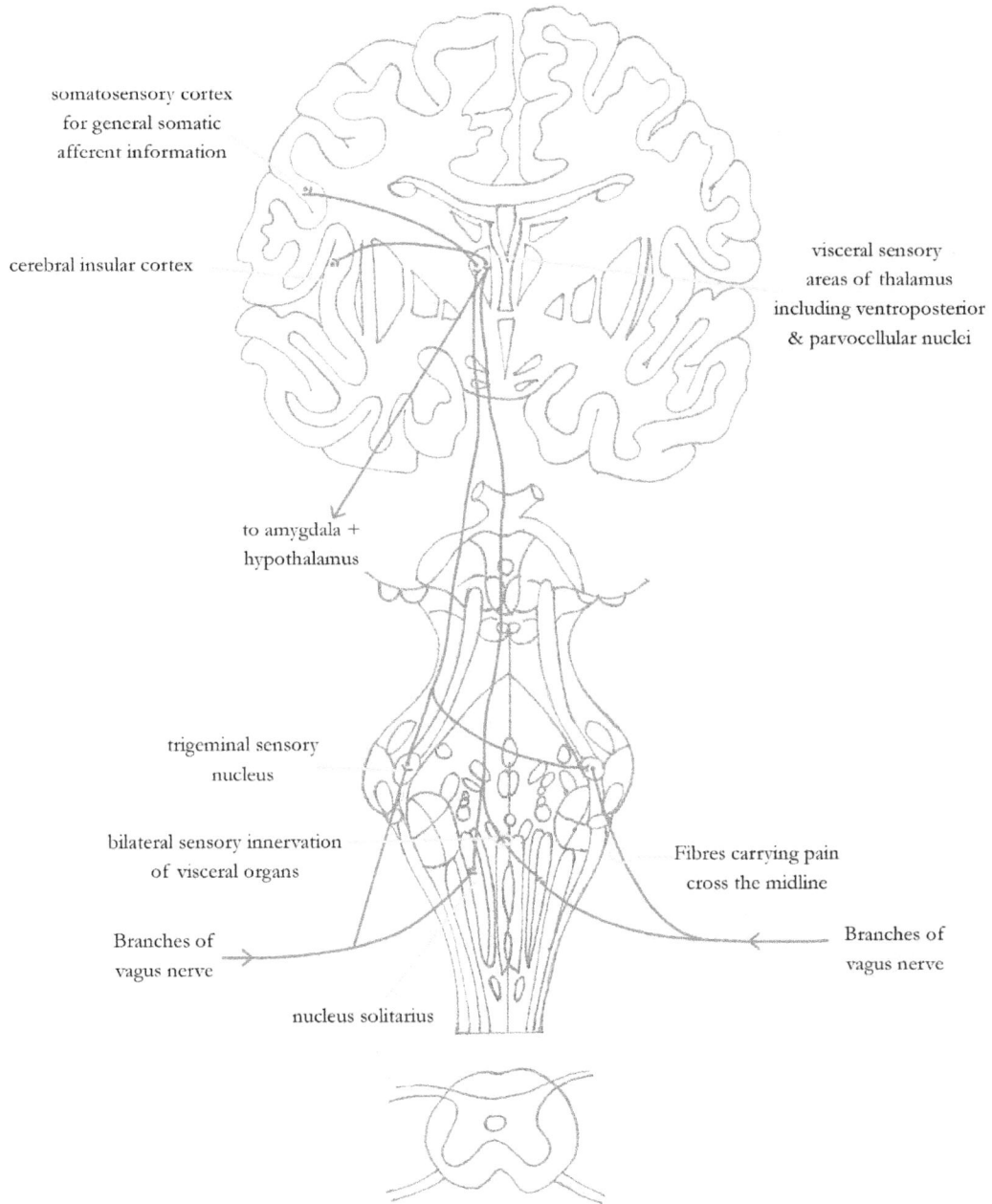

somatosensory cortex
for general somatic
afferent information

cerebral insular cortex

visceral sensory
areas of thalamus
including ventroposterior
& parvocellular nuclei

to amygdala +
hypothalamus

trigeminal sensory
nucleus

bilateral sensory innervation
of visceral organs

Fibres carrying pain
cross the midline

Branches of
vagus nerve

Branches of
vagus nerve

nucleus solitarius

Vagus Nerve CNX Branchial Motor Pathway

Special visceral efferent

PRIMARY MOTOR CORTEX
(Upper motor neuron)

NUCLEUS AMBIGUUS
(Synapse to lower motor neuron; fibres descend
ipsilaterally and contalaterally)

VAGUS NERVE
(Motor fibres for speech & swallow – project from
nucleus ambiguus)

BRANCHES OF VAGUS NERVE

MUSCLES OF THE VELUM, PALATOGLOSSUS MUSCLE SUPERIOR/
MIDDLE/INFERIOR PHARYNGEAL CONTRICTOR MUSCLES,
INTRINSIC MUSCLES OF LARYNX

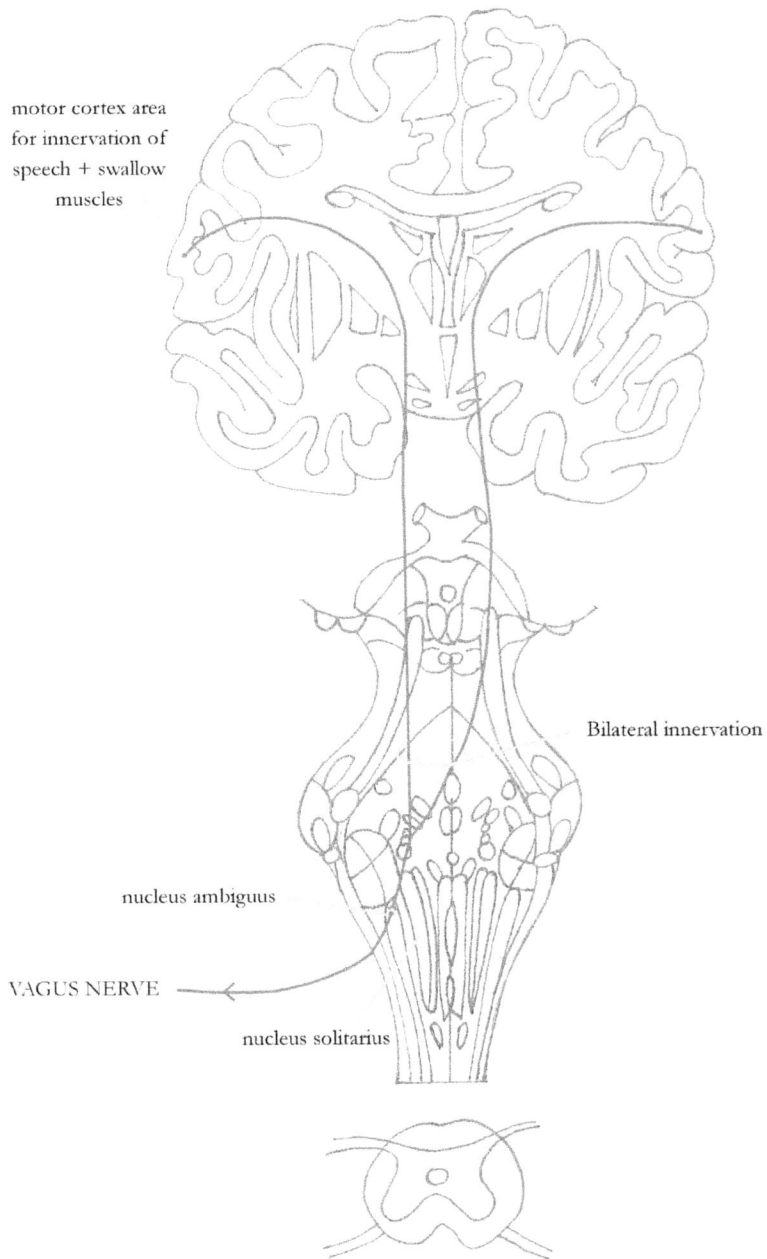

motor cortex area
for innervation of
speech + swallow
muscles

Bilateral innervation

nucleus ambiguus

VAGUS NERVE

nucleus solitarius

Vagus Nerve CNX Parasympathetic Pathway

General visceral efferent

HYPOTHALAMUS & AMYGDALA

DORSAL VAGAL NUCLEI
(Fibres descend ipsilaterally)

NUCLEUS OF SOLITARY TRACT

VAGUS NERVE

DIVISIONS OF VAGUS NERVE

VISCERAL ORGANS
(RESPIRATORY, CARDIOVASCULAR, GASTROINTESTINAL, PELVIC)

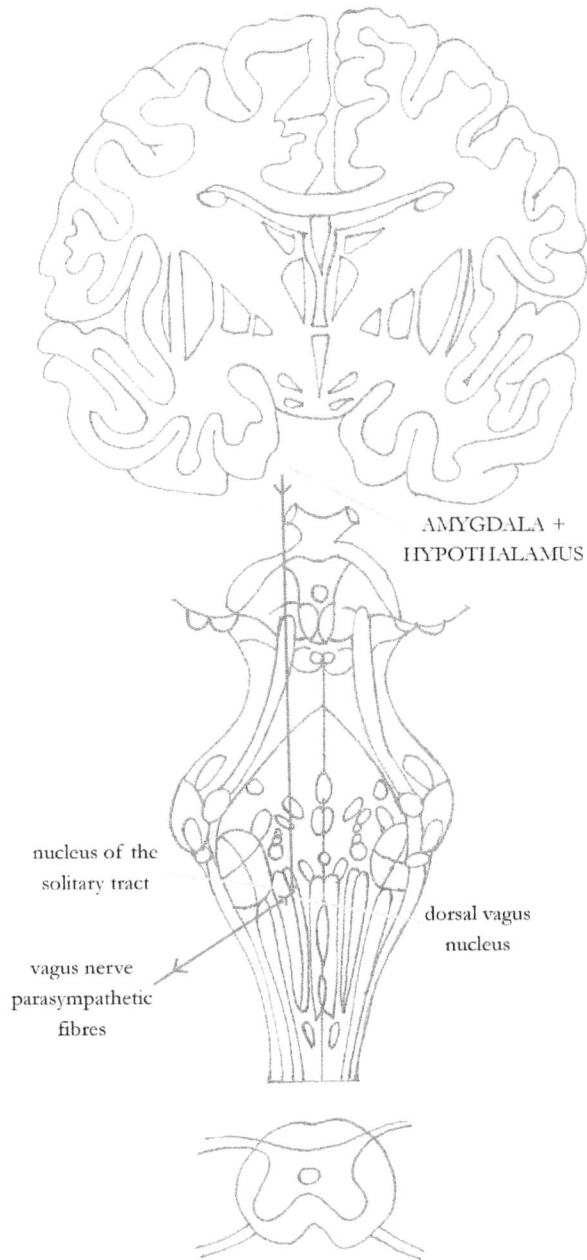

AMYGDALA +
HYPOTHALAMUS

nucleus of the
solitary tract

dorsal vagus
nucleus

vagus nerve
parasympathetic
fibres

Accessory Nerve CNXI

Branchial motor innervation of sternocleidomastoid & trapezius muscles
General somatic efferent

PRIMARY MOTOR CORTEX
(Upper motor neuron)

NUCLEUS AMBIGUUS OF MEDULLA OBLONGATA
& SPINAL GREY MATTER C1-C5
(Synapse to lower motor neuron; fibres descend ipsilaterally to
sternocleidomastoid and contralateral to trapezius)

ROOTLETS OF ACCESSORY NERVE

ACCESSORY NERVE

NEUROMUSCULAR JUNCTION OF STERNOCLEIDOMASTOID
& TRAPEZIUS MUSCLES

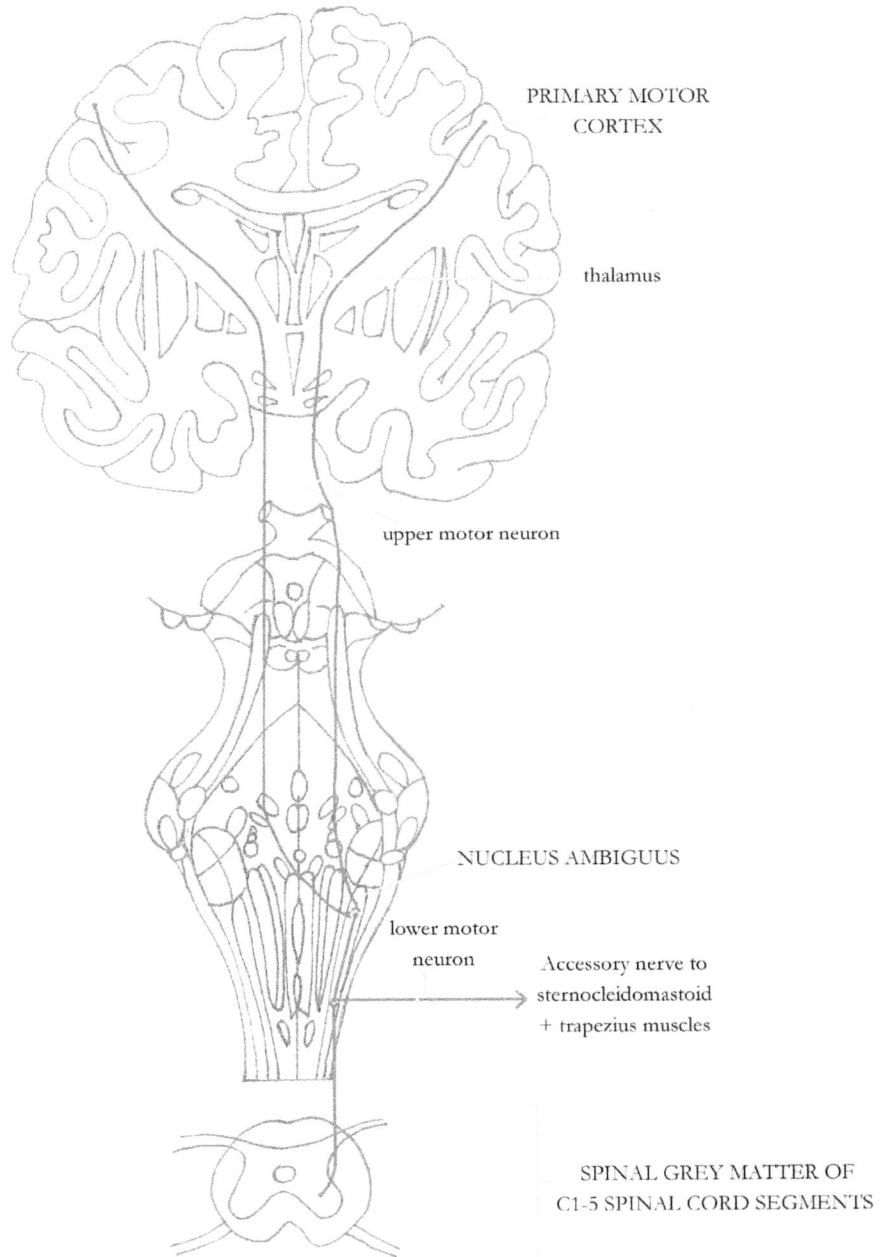

PRIMARY MOTOR
CORTEX

thalamus

upper motor neuron

NUCLEUS AMBIGUUS

lower motor
neuron

Accessory nerve to
sternocleidomastoid
+ trapezius muscles

SPINAL GREY MATTER OF
C1-5 SPINAL CORD SEGMENTS

Hypoglossal Nerve CNXII

Somatic motor innervation of extrinsic & intrinsic muscles of the tongue
General somatic efferent

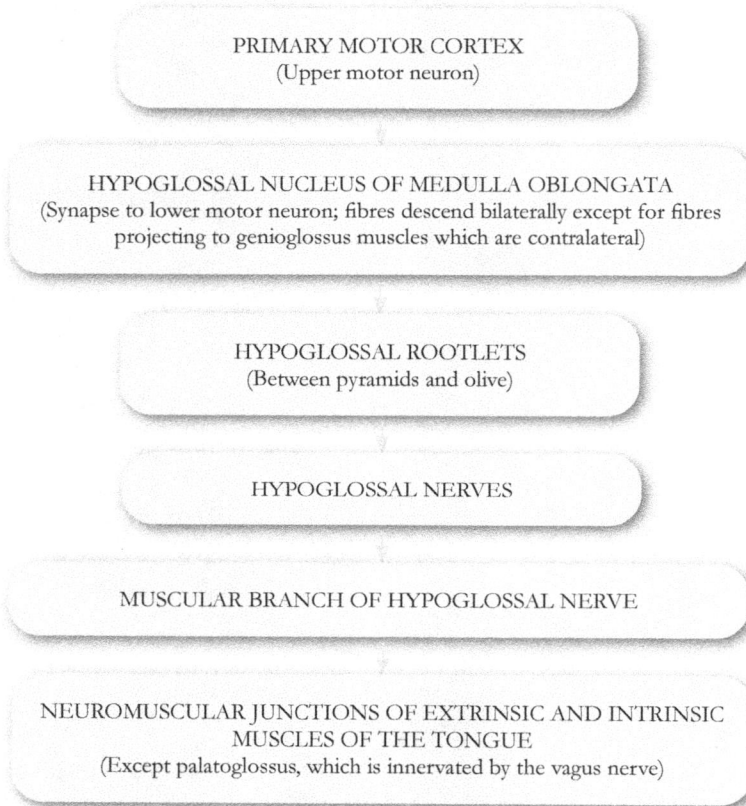

PRIMARY MOTOR CORTEX
(Upper motor neuron)

HYPOGLOSSAL NUCLEUS OF MEDULLA OBLONGATA
(Synapse to lower motor neuron; fibres descend bilaterally except for fibres
projecting to genioglossus muscles which are contralateral)

HYPOGLOSSAL ROOTLETS
(Between pyramids and olive)

HYPOGLOSSAL NERVES

MUSCULAR BRANCH OF HYPOGLOSSAL NERVE

NEUROMUSCULAR JUNCTIONS OF EXTRINSIC AND INTRINSIC
MUSCLES OF THE TONGUE
(Except palatoglossus, which is innervated by the vagus nerve)

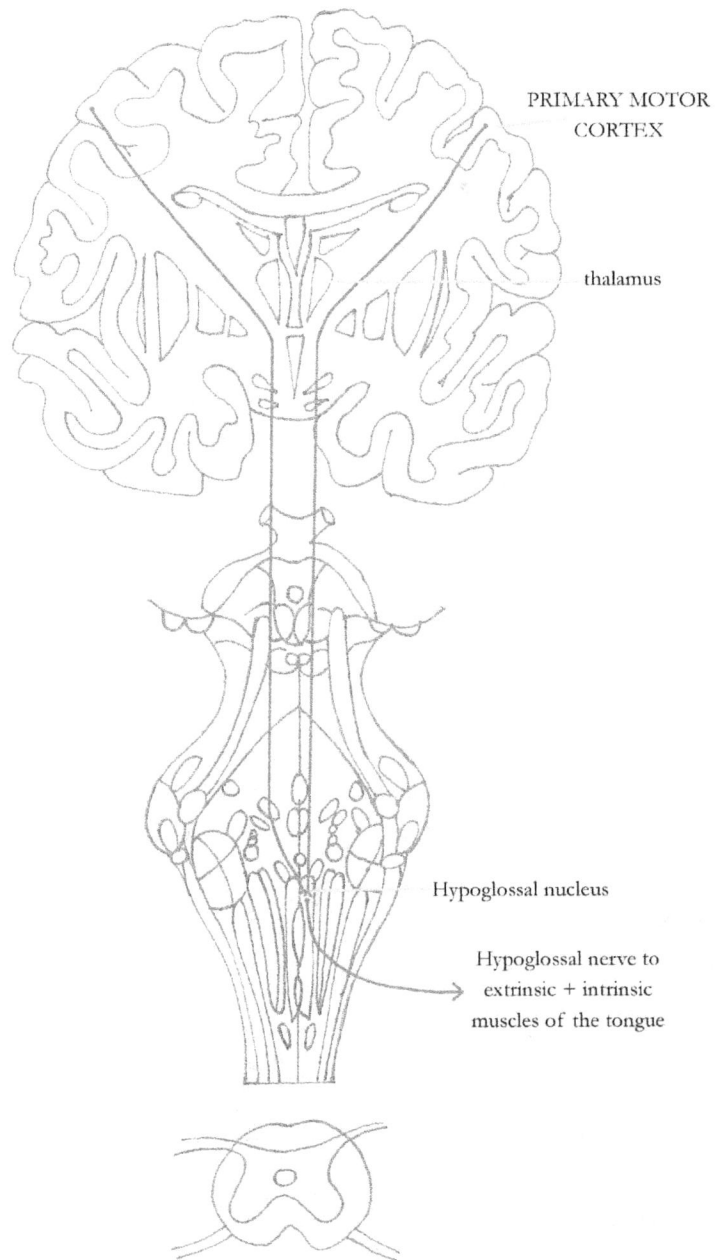

PRIMARY MOTOR
CORTEX

thalamus

Hypoglossal nucleus

Hypoglossal nerve to
extrinsic + intrinsic
muscles of the tongue

REFERENCES

Allen, G. V., & Cechetto, D. F. (1994). Serotoninergic and nonserotoninergic neurons in the medullary raphe system have axon collateral projections to autonomic and somatic cell groups in the medulla and spinal cord. *The Journal of Comparative Neurology, 350*(3), 357-366. https://doi.org/10.1002/cne.903500303

Bear, M. H., Reddy, V., & Bollu, P. C. (2021). *Neuroanatomy, Hypothalamus.* StatPearls Publishing LLC.

Bradley, R. M. (2007). Chapter 1, Historical Perspectives. In R. M. Bradley (Eds.), *The role of the nucleus of the solitary tract in gustatory processing.* (pp. 1-20). Taylor & Francis Group, LLC. List of Contributors - The Role of the Nucleus of the Solitary Tract in Gustatory Processing - NCBI Bookshelf (nih.gov)

Breit, S., Kupferberg, A., Rogler, G & Hasler, G. (2018). Vagus nerve as modulator of the brain-gut axis in psychiatric and inflammatory disorders. *Frontiers in Psychiatry, 9*(44), 1-15. DOI: 10.3389/fpsyt.2018.00044

Chason, H. M., & Downs, B. W. (2018). *Anatomy, Head and Neck, Parotid Gland.* StatPearls Publishing LLC.

Cacciola, A., Milardi, D., Basile, G. A., Bertino, S., Calamuneri, A., Chillemi, G., Paladina, G., Impellizzeri, F., Trimarchi, F., Anastasi, G., Bramanti, A., & Rizzo, G. (2019). The cortico-rubral and cerebello-rubral pathways are topographically organized within the human red nucleus. *Nature Reseaerch, 9*(12117), 1-12. https://doi.org/10.1038/s41598-019-48164-7

Coote, J. H., Spyer, K. M. (2018). Central control of autonomic function. *Brain and Neuroscience Advances, 2,* 1-5. DOI: 10.1177/2398212818812012.

Crespo, C., Liberia, T., Blasco-Ibanez, J. M., Nacher, J., & Varea, E. (2018). Cranial pair I: The olfactory nerve. *The Anatomical Record, 302,* 405-427. Doi: 10.1002/ar.23816

Crossman, A. R., & Neary, D. (2000). *Neuroanatomy: An Illustrated Colour Text.* (2nd ed.). Philadelphia, USA: Elsevier Limited

de Castro, D. C., & Marrone, L. C. (2021). *Neuroanatomy, Geniculate Ganglion.* StatPearls Publishing LLC.

Dulak, D & Naqvi, I. A. (2020). *Neuroanatomy, cranial nerve 7 (facial).* StatPearls Publishing LLC.

Fukushima, K., Hirai, N., & Rapoport, S. (1979). Direct excitation of neck flexor motor neurons by the interstatiospinal tract. *Brain Research, 160,* 358-362. https://doi.org/10.1016/0006-8993(79)90432-3

Fuller, D. R., Pimentel, J. T., & Peregoy, B. M. (2012). Applied Anatomy and Physiology for Speech-Language Pathology and Audiology. Philadelphia, USA: Lippincott Williams & Wilkins.

Gao, Y., & Sun, T. (2016). Molecular regulation of hypothalamic development and physiological functions. *Molecular Neurobiology, 53*(7), 4275-4285. DOI: 10.1007/s12035-015-9367-z

Garcia Santos, J. M., Sanchez Jimenez, S., Tovar Perez, M., Moreno Cascales, M., Laihacar Marty, J., & Fernandez-Villacanas Marin, M. A. (2018). Tracking the glossopharyngeal nerve pathway through anatomical references in cross-sectional imaging techniques: A pictorial review. *Insights Imaging, 9*(4), 559-569. DOI: 10.1007/s13244-018-0630-5

Gibbons, J. R., & Sadiq, NM. (2021). *Neuroanatomy, Neural Taste Pathway.* StatPearls Publishing LLC.

Gillig, P. M., & Sanders, R. D. (2010). Cranial Nerves IX, X, XI, and XII. *Psychiatry, 7*(5), 37-41.

Giraudin, A., Cabirol-Pol, M., Simmers, J., & Morin, D. (2008). Intercostal and abdominal respiratory motoneurons in the neonatal rat spinal cord: Spatiotemporal organization and responses to limb afferent stimulation. *Journal of Neurophysiology, 99,* 2626-2640. DOI: 10.1152/jn.01298.2007.

Haines, D. E. (2004). *Neuroanatomy: an atlas of structures, sections, and systems.* Lippincott Williams & Wilkins.

Kandel, E. R., Schwartz, J. H., & Jessell, T.M. (2000). *Principles of Neural Science.* (4th ed.). USA: McGraw-Hill Companies

Kenny, B. J., & Bordoni, B. (2021). *Neuroanatomy, Cranial Nerve 10 (Vagus Nerve).* StatPearls Publishing LLC.

Kiernan, J. A. (1998). *Barr's The Human Nervous System: An Anatomical Viewpoint.* (7th ed.). Philadelphia, USA: Lippincott-Raven Publishers

Kim, S. Y., & Naqvi, I. A. (2020). *Neuroanatomy, Cranial Nerve 12 (Hypoglossal).* StatPearls Publishing LLC.

Klingner, C. M., & Witte, O. W. (2015). Central Facial Palsy. In O. Guntinas-Lichius & B. M. Schaitkin, *Facial Nerve Disorders and Diseases: Diagnosis and Management.* (pp. 357-359). Thieme Medical Publishers, Incorporated.

Larson, B., Miller, S., & Oscarsson, O. (1969). Termination and functional organisation of dorsolateral spino-olivocerebellar path. *Journal of Physiology, 203,* 611-640. DOI: 10.1113/jphysiol.1969.sp008882

Loewy, A. D. (1982). Descending pathways to the sympathetic preganglionic neurons. *Progress in Brain Research, 57,* 267-277. https://doi.org/10.1016/S0079-6123(08)64133-3

Mtui, E. P., Anwar, M., Gomez, R., Reis, D. J., & Ruggiero, D. A. (1993). Projections from the nucleus tractus solitarii to the spinal cord. *The Journal of Comparative Neurology, 337,* 231-252. https://doi-org. libraryproxy.griffith.edu.au/10.1002/cne.903370205

Mtui, E., Gruener, G., & Dockery, P. (2016). *Fitzgerald's Clinical Neuroanatomy and Neuroscience.* (7th ed.).

Nayagam, B. A., Muniak, M. A., & Ryugo, D. K. (2011). The spiral ganglion: connecting the peripheral and central auditory systems. *Hearing Research, 278*(1-2), 2-20. DOI: 10.1016/j.heares.2011.04.003

Patel, NM., & Das J, M. (2020). *Neuroanatomy, Spinal Trigeminal Nucleus.* StatPearls Publishing LLC.

Putnam, S. J., & Manning, J. W. (1977). Repetitively firing medullary neurons responsive to carotid sinus nerve stimulation and norepinephrine infusion. *Brain Research, 122,* 556-561. https://doi. org/10.1016/0006-8993(77)90467-X

Qiu, K., Lane, M. A., Lee, K. Z., Reier, P. J., & Fuller, D. D. (2010). The phrenic motor nucleus in the adult mouse. *Experimental Neurology, 226*(1), 254-258. DOI: 10.1016/j.expneurol.2010.08.026.

Samuels, E. R., & Szabadi, E. (2008). Functional neuroanatomy of the noradrenergic locus coeruleus: Its roles in the regulation of arousal and autonomic function part 1: Principals of functional organization. *Current Neuropharmacology, 6*(3), 235-253. DOI: 10.2174/157015908785777229

Santos, J. M. G., Jimenez, S. S., Perez, M. T., Cascales, M. M., Marty, J. L., & Fernandez-Villacanas Marin, M. A. (2018). Tracking the glossopharyngeal nerve pathway through anatomical references in cross-sectional imaging techniques: a pictorial review. *Insights into Imaging, 9*, 559-569. DOI: 10.1007/s13244-018-0630-5

Selvaraj, K., Gowthamarajan, K., & Venkata Satyanarayana Reddy Karri, V. (2017). Nose to brain transport pathways an overview: potential of nanostructured lipid carriers in nose to brain targeting. *Artificial Cells, Nanomedicine, and Biotechnology, 46*(8), 2088-2095. DOI: 10.1080/21691401.2017.1420073

Seneviratne, S. O., & Patel, B. C. (2020). *Facial nerve anatomy and clinical applications.* StatPearls Publishing LLC.

Shankland, W. E. (2001). The trigeminal nerve. Part II: The ophthalmic division. *The Journal of Craniomandibular and Sleep Practice, 19*(1), 8-12. DOI: 10.1080/08869634.2001.11746145

Slattery, W. H., & Azizzadeh, B. (2014). *The Facial Nerve.* ProQuest Ebook Central. https://ebookcentral-proquest-com.libraryproxy.griffith.edu.au/lib/griffith/reader.action?docID=5252892

Sonne, J., & Lopez-Ojeda, W. (2020). *Neuroanatomy, Cranial Nerve.* StatPearls Publishing LLC.

Standring, S., Ellis, H., Healy, J. C., Johnson, D., Williams, A., Collins, P., Wigley, C., Berkovitz, B. K. B. … Shah, P. (2005). *Gray's Anatomy: The Anatomical Basis of Clinical Practice.* (2nd ed.). Philadelphia, USA: Elsevier Ltd

Thomas, K., Minutello, K., & Das, J. M. (2020). *Neuroanatomy, Cranial Nerve 9 (Glossopharyngeal).* StatPearls Publishing LLC.

Verberne, A. J. M. (2003). Medulla oblongata: Glossopharyngeal nerve (Cranial nerve IX). In M. J. Aminoff & R. B. Daroff, *Encyclopaedia of the Neurological Sciences* (pp. 54-63). Academic Press.

Wilson-Pauwels, L., Stewart, P. A., Akesson, E. J., & Spacey, S. D. (2010). *Cranial Nerves: Function & Dysfunction.* (3rd ed.). Elsevier Australia.

Yolas, C., Kanat, A., Aydin, M. D., Turkmenoglu, O. N., & Gundogdu, C. (2014). Important casual association of carotid body and glossopharyngeal nerve and lung following experimental subarachnoid hemorrhage in rabbits. First report. *Journal of Neurological Sciences, 336*(1), 220-226. (Griffith university library https://www-clinicalkey-com-au.libraryproxy.griffith.edu.au/#!/content/1-s2.0-S0022510X13030098

www.ingramcontent.com/pod-product-compliance
Lightning Source LLC
Chambersburg PA
CBHW051226200326
41519CB00025B/7268